My STEAM Notebook

Helping Kids Write About Their Observations

by
Darcy Pattison

Foreword by
Karen Ansberry and Emily Morgan

Mims House | Little Rock

© 2017 Darcy Pattison
ISBN: 978-1-62944-072-9

All Rights Reserved. No part of this publication may be reproduced, distributed or transmitted in any form or by any means, including photocopying, recording, or other electronic or mechanical methods, without the prior written permission of the publisher, except in the case of brief quotations embodied in critical reviews and certain other noncommercial uses permitted by copyright law.

Mims House
1309 Broadway
Little Rock, AR 72202
MimsHouse.com

Publisher's Cataloging-in-Publication data

Names: Pattison, Darcy, author.
Title: My STEAM notebook : Helping Kids Write About Their Observations
by Darcy Pattison ; foreword by Karen Ansberry and Emily Morgan
Description: Little Rock, AR : Mim's House, 2017
Identifiers: ISBN 978-1-62944-072-9 | LCCN 2016914531
Subjects: LCSH Science--Methodology--Juvenile literature. | Notebooks--Juvenile literature. | Scientists--Juvenile literature. | Science--Study and teaching (Elementary)--United States. | BISAC JUVENILE NONFICTION / Science & Nature / General
Classification: LCC Q175.2 .P37 2016 | DDC 509.2/2--dc23

Foreword

We had the pleasure of meeting author Darcy Pattison in the spring of 2016 at a Picture-Perfect Science teacher workshop we were facilitating in Arkansas. We were very familiar with Darcy's work, as we had just written a lesson for our new book, *Picture-Perfect STEM Lessons*, using one of her latest nonfiction picture books for children: *Burn: Michael Faraday's Candle*. When we found out that Darcy lived near the workshop location, we were delighted that she would be able to join us one afternoon to read some of her books and talk with teachers about her writing process. There she read aloud *Nefertiti the Spidernaut: The Jumping Spider Who Learned to Hunt in Space*, and this remarkable true story inspired us to write a "Spider Science" lesson to include in our new book. Afterwards, Darcy joined us for dinner and we had a lively discussion about interesting topics for nonfiction books, the state of elementary science education in general, and the types of curricular resources that might help elementary science teachers better integrate science, reading, and writing. We all acknowledged the struggle that many teachers across the country face to include science and engineering in the curriculum, when accountability pressures for student achievement in reading, writing, and mathematics are ever increasing. During that conversation, the concept of a book to help teachers implement STEM concepts through notebooking was discussed. Darcy embraced the idea and ran with it! The result is *My STEAM Notebook*, an intriguing look at the notebooks of several great American scientists, done in an engaging, interactive format.

Teachers and parents alike will appreciate this kid-friendly approach to science notebooking. Take a look inside, and you will find original source documents to help children understand the multidisciplinary nature of the work that scientists do. Reading through these historical documents, along with the biographical information about each scientist, clearly shows how scientists use writing, drawing, mathematics, and problem solving as they seek to understand the natural world.

This book's innovative format not only lets students peek inside the notebooks of actual scientists, but invites them to record their own ideas right alongside. The book's process-oriented approach provides students with real-world examples of science and engineering practices, opportunities to engage in these practices for themselves, and a place to record their experiences.

We have no doubt that *My STEAM Notebook* will be an invaluable resource for teachers who want their students to experience the connections among science, technology, engineering, art, and mathematics in a uniquely personal way.

Karen Ansberry and Emily Morgan
Cofounders, Picture-Perfect Science
Authors of the *Picture-Perfect Science Lessons* series (NSTA Press)

For Parents and Educators
Why a Science Notebook?

"One item which seems quite naturally to be inseparable from you in your work is your notebook. Many years ago, during my Congo wanderings, I was given a ring file pocketbook, which has been a treasured and useful memento of that country for more than three decades. . . --Jack Vincent, British ornithologist" [1]

Scientists tend to be fanatical about their notebooks. For those who get into the habit of recording in a notebook, it becomes a confidante. It includes their thoughts, actions, evaluations, dreams, speculations, observations, tedious lists of specimens, and much more. It's fitting that young scientists turn to these historical journals for clues on what to include in their own science notebooks.

Observing historical science notebooks

To write this book, I looked at hundreds of different notebooks from a variety of American scientists. Most came from the Smithsonian Field Book project[2] and the National Library of Medicine[3]. Notebooks from biologists and doctors are different. Throw in the notebooks from the Silicon Valley engineers housed at the Computer History Museum[4], and scientists' notebooks expressed many different goals and approaches. Some emphasized one step of the scientific process more than another. Each notebook looks different because scientists were trying to accomplish different goals. Even the shapes of the physical books varied.

Engineers tended to emphasize idea generation, the design phase, or drawings of how to build something. Biologists tended to tell a narrative of observing or collecting specimens in the wild. In the laboratory, notebooks tended to be more procedural, or "this is what I did and how I did it." Medical research included be exact chemical procedures in a laboratory. Notebooks for those researchers held pages of mathematical figures, dense tables of data, and little narrative. Doctors involved in public health, however, traveled to sites with disease outbreaks, worked with community organizers to make changes, or worked on public education campaigns. Their notebooks are often travelogues with notes on disease scattered throughout.

Some scientists were compulsive about writing down everything, while others merely jotted things now and then. Overseas travel often inspired a detailed diary, and then the scientist wrote nothing for a decade. But through the varied experiences of American scientists, the notebooks are there. Why?

Scientists felt compelled to keep a notebook for many reasons. For engineers, a notebook could be a legal document, the basis of a patent filing. Other scientists seemed to have a sense of destiny and wanted to record something for later generations to read. Others were just bugged by an idea and wanted to work it out on paper.

Essentially, they all had to address the basic question of all writing: who is your audience? Yourself or others?

Process v. Product based Notebooks

Most notebooks I looked at took a process-based approach, which means the notebook was a record of the process of exploring science. These notebooks were written by the scientists for themselves. Even when there was a sense that this record might be historically important, scientists often skipped days in recording data.

By contrast, most recommendations about student science notebooks take a product-based approach. Students must complete a project with certain required elements, and the teacher grades the notebook. Scientists are focused inward on their own goals, experiences, and projects. Students, because they produce a product-based notebook, must look outward. Scientists write for themselves; students write for their teacher. Like any writing project, audience is a key consideration of what and how something is written.

One element almost universally required in student notebooks is a question. Often called a focusing question, it serves to guide the rest of the inquiry. After examining historical examples of notebooks from scientists, I rarely found a focusing question. That's not to say that the question wasn't in the scientist's mind, but it wasn't expressed on the pages of notebooks. Scientists were usually clear in their inquiry goals and didn't need to state the question so others could evaluate it. Again, it's the difference between inward or outward facing purposes for a notebook.

Another way to say this is that process-based notebooks are best used for formative assessment, those which monitor student understanding and then modify the course work to aid understanding. Product-based science notebooks are best for summative assessment such as when the teacher evaluates and assigns a grade.

150 Years of American Scientists

The scientists whose notebooks are included here span about 150 years of American scientific study, from the mid-1800s to the end of the 1900s. In the process of researching available historical notebooks, I concentrated on seeking examples that would help students learn to use their own notebooks to record questions, observations, and conclusions. The historical notebooks are arranged here in a progression that will help students understand the potential for what a notebook can do for their scientific understanding.

Alexander Wetmore, nicknamed Alick (pp. 16-17), is presented first because his first recording of a bird occurred at age eight while in Florida on a vacation. He described the pelican as a "great big bird that eats fish."[5] Throughout his teen years, he kept a monthly record of all the birds he saw. By age 15, he had published his first article in 1900 in *Bird Lore* magazine, "My Experience with a Red-headed Woodpecker." (See pp. 148-149 for a reproduction of that article.) Wetmore's notebooks show that observations can be done at any age. Lifelong passions can begin in an elementary school science notebook.

If you pare it down to essentials, the only things recorded in a notebook are words and drawings. Of course, photographs, worksheets, or other memorabilia can be fastened inside the notebook, but what students will actually write are words and drawings. Students need to explore a variety of ways to use text and art.

Martin H. Moynihan (pp. 28-29) presents a variety of options: text only, drawings only and a combination of text and drawing. Sometimes, text dominates, and other times drawings dominate.

Likewise, William Healey Dall (pp. 40-41) gives students a look at additional options possible in a notebook. He drew maps, native people, and interesting objects while he kept a careful record of his travels to Alaska. Look especially at his drawing of native pottery. While it's interesting, the drawing alone doesn't tell enough because we don't know the scale. Only the text explains the size of each pot. Students need to learn to use text and drawings together to give a more complete understanding of what is observed.

A basic skill that students need is the ability to make a careful observation. Joseph Nelson

Rose's cactus example (pp. 52-53) is excellent because he includes descriptions of color, size, shape, and number. Notice too that he uses scientific vocabulary. As students write in notebooks, observations will be more exact as they learn the scientific names for objects, anatomy, and so on. For that, use My Glossary in the back of this book. However, remember that students may also choose to define words in context.

Lucile Mann (pp. 64-65) was the wordsmith in the family, leaving the public speaking to her husband, William "Bill" Mann, Director of the National Zoo. Because she worked first as an editor, her diaries are carefully typed and edited. One type of writing found over and over in science notebooks is a narrative, or a description of something that happened to them. Mann's narrative writing skills are shown by her use of sensory details in her travel descriptions.

Fred Soper (pp. 76-77) also recorded narratives in his diaries kept during public health work in Brazil. He not only records scientific observations, but does it with humor. His writing voice was warm, sarcastic and funny.

Shifting focus to the drawings, several scientists were especially adept at sketching.

Mary Agnes Chase (pp. 88-89) originally worked as a botanical illustrator. Early in her career, she learned to use a microscope which helped her make observations that brought her work to life. She also used photography extensively later in her career, and it's interesting to discuss with students the role of a botanical illustrator as compared with a photographer. Illustrators are free to combine elements from different seasons: for example a flower and a fruit. Photographers are restricted to only what their cameras can record. Also look at how carefully her type-written pages are edited.

While many of the scientists included drawings, Donald S. Erdman (pp. 100-101) took them to a new level with color (although shown in b/w here). But he didn't use color just to use color. Instead, he describes the reason for color: that preserved fish quickly lose any color. For proper identification and understanding of the fish, color was required. Students should learn to use whatever tools are necessary to record observations.

Robert E. Silberglied (pp. 112-113) had an amazing eye for visual details. Notice the elaborate key and compass indicating north that he used on his map of Gomez Farias in Mexico. Silberglied also specialized in photography. He used ultraviolet light in his studies and photographed flowers in ultraviolet light. Optical microscopy allowed him to zoom in close on a butterfly's wing. Though he didn't use it, we introduce the idea of aerial or satellite photography and electron microscopy in the discussion questions.

Almost all these American scientists collected specimens. Throughout, you'll see discussions of objects that are sent back home for further study. From Chase's grasses to Wetmore's bird skins, collecting items for further study is an important part of observation. Scientists were careful to record exactly when and where the items were collected. Often the descriptions involve a physical location (e.g. Silberglied's ". . .2 miles off Mexican Highway 85"[6]) Temperature, weather, elevation and other conditions are often reported. Students need to learn to record these type of variables.

Watson M. Perrygo (pp. 124-125), as a taxidermist and museum curator, shows one of the final stages of observations and collection of specimens. The objects are available for various scientific studies, and they are also made available for the general public to view in a museum setting. The specimens are important historical snapshots of an ecosystem and can be compared to contemporary conditions. But they are also an entertaining way to learn more science. Museums write informational materials to help the public understand what they are seeing.

Science and History

While these historical notebooks are presented for their value in learning to study science, they also are historical documents. The scientists lived and worked at particular times, and their experiences are tied to those time periods.

William Dall traveled with the Western Telegraph Expedition to Alaska, known in 1867 as Russian America. While on his trip, on March 30, 1867, America bought Alaska from Russia. They paid $7.2 million. It took several months for word of the purchase to reach the Expedition in Alaska.

Lucile and Bill Mann went to Sumatra (Indonesia) in 1937. They stopped in Tokyo, and on February 5, 1937, Japanese Prince Taka Tsucasa (Takatsukasa Nobusuke) showed them his aviary. The Prince later authored a book about the birds of Japan. After viewing the "paroquets, pigeons and pheasants,"[7] the Manns joined the Prince in a formal Japanese meal. Just four years later, on December 7, 1941, Japan bombed Pearl Harbor.

Donald S. Erdman spent World War II in the Civilian Public Service because he was a conscientious objector.

In 1965, Robert E. "Bob" Silberglied traveled at age 19 on a Cornell University Ant Safari, a trip to collect ants in Mexico. They traveled from New York to Mexico, passing through Mississippi. In March that year, Civil Rights activists marched in Selma, Alabama, protesting the death of protesters. When the Cornell Ant Safari people passed through Mississippi on July 12, Bob wrote: "At stop for gas in Sandersville, Mississippi, there was a Ku Klux Klan Poster on a tree. I removed it before we left. It reads: 'I want you (Picture of Uncle Sam in KKK uniform, with pointing figure) in the middle of White Knights of Mississippi Ku Klux Klan (colors: red, white, & blue)'."[8]

Each scientist featured has an interesting story partly because of the time period in which he or she lived. The focus of this notebook isn't on the historical events of their times, but on their scientific pursuits. The culture, technology, engineering, world events and much more influenced what was possible for each of them.

Steam + History

The scientists are arranged topically, so that the progression leads to a stronger understanding of what is possible in a science notebook. However, if you wish to study them in historical context, here's the list of scientists in chronological order along with an important world event that affected their lives.

William Healey Dall 1845-1927, purchase of Alaska from Russia
Joseph Nelson Rose 1862-1928 World War I
Mary Agnes Chase 1869-1963 Suffragist Movement, 19th Amendment
Alexander Wetmore 1886-1978 Panama Canal
William "Bill" Mann 1886-1960 World War I and II
Fred Soper 1893-1977 Typhus and Yellow Fever breakouts in World War II
Lucille Mann 1897-1986 World War I and II
Watson M. Perrygo 1906-1984 Panama Canal
Donald S. Erdman 1919- ? World War II conscientious objector, Persian Gulf
Martin H. Moynihan 1928-1996 Panama Canal
Robert E. "Bob" Silberglied 1946-1982 Civil Rights, Vietnam War

How to Use This Notebook

S T E A M This book supports the Science-Technology-Engineering-Art-Math initiative with some unique features. Throughout the text, you'll see these letters: S T E A M. This indicates a related discussion questions for each scientist's work at the back of the book. The questions will start discussions relating to the different disciplines, but of course, the discussions can range much farther.

The Table of Contents should be used every day. Here, students will summarize the day's work, who helped and the page numbers used in the notebook. The layout of the Table of Contents and open-ended pages is important. Numbered pages make the Table of Contents easy to maintain. The headings on the notebook pages are deceptively simple, but they encourage consistent recording of potentially important information.

The pages are open-ended, with space for writing and drawing. Graph paper is provided on the odd-numbered pages for math work or to aid handwriting. Even-numbered pages are blank for drawing, writing, or adding worksheets. Using this open-ended notebook will inevitably mean some students have messy pages that reflect the uneven nature of observations. Organization may be sketchy, other than the Table of Contents. In other words, they will mirror the actual work of scientists who are process-oriented in their work.

However, the reality of education is that students sometimes work in a product-based notebook for summative assessment. For those times, you can easily glue in worksheets as needed. We recommend no staples or glue sticks; instead, use liquid glue that's more permanent. Markers often bleed through pages, so colored pencils or crayons are recommended.

The reading level of the student sections of this notebook is third grade, within the 2nd-3rd grade Lexile bands of 420L-820L.

Scientific Arguments - Claim, Evidence and Explanation

Science and writing standards today emphasize the need to argue from evidence. The practice of observation in notebooks provides evidence, but students need help in turning that information into a coherent argument. Educators Lori Fulton and Emily Poeltler explain some factors that help.[9] Sentence frames give students the language they need to think about the bigger ideas of their observations.

Sentence frames for claims include:
- I think. . .
- I observed. . .
- I noticed. . .
- A/an. . . is an example of. . .

For evidence, students can use these sentence frames:
- I found. . .
- My evidence is . . .
- My reasons are. . .

Here's an explanation sentence frame:
   This happened because...

Using sentence frames to develop talking points is only the first step. Students need discussion time to talk through ideas, and they often need modeling to learn to argue from evidence. Still, frames are a starting point. *My STEAM Notebook* includes IDEAS TO TALK ABOUT writing prompts with some of these sentence frames in each section. Use them as needed.

THIS NOTEBOOK BELONGS TO:

TO THE STUDENTS:
When you finish a day's work, summarize the work, who helped and the page number used.

TABLE OF CONTENTS

TITLE OR SUBJECT	INVESTIGATOR(S)	DATE	PAGE(S)

THIS NOTEBOOK BELONGS TO:

TABLE OF CONTENTS

TITLE OR SUBJECT	INVESTIGATOR(S)	DATE	PAGE(S)

THIS NOTEBOOK BELONGS TO:

TABLE OF CONTENTS

TITLE OR SUBJECT	INVESTIGATOR(S)	DATE	PAGE(S)

THIS NOTEBOOK BELONGS TO:

TABLE OF CONTENTS

TITLE OR SUBJECT	INVESTIGATOR(S)	DATE	PAGE(S)

THIS NOTEBOOK BELONGS TO:

TABLE OF CONTENTS

TITLE OR SUBJECT	INVESTIGATOR(S)	DATE	PAGE(S)

Alexander Wetmore

1886 - 1978 ornithologist or bird scientist
Birthplace: North Freedom, Wisconsin
Education: B.S. 1912 University of Kansas, M.S. 1916, and Ph.D. 1920 George Washington University

Alexander Wetmore in Panama, 1953.

Alexander Wetmore started writing in his journals at 8-years-old. His family went to Florida for vacation. He saw a pelican and wrote about it. By age 14, he was writing monthly lists of birds that he saw.

Wetmore's first job was for the U.S. Bureau of Biological Survey. He studied birds in Latin America and Puerto Rico (1911). For two years, he traveled in South America. He studied bird migration from North America to South America. Later, he studied the food habits of North American birds.

Wetmore worked for the Smithsonian Museum from 1925-1952. He served as the sixth Secretary from 1945-1952. During the Great Depression of the 1930s, he

Journaling at Any Age

In 1895, at the age of 8, Wetmore wrote in a journal. His family was in Florida on vacation. He wrote a list of things he saw one day:

"pelicans, fish, shark, stingray, air gun, pressed flowers. There are a great many pelicans around here. A pelican is a great big bird that eats fish." S A

This image shows Wetmore's February, 1904 journal, age 18. He lists 24 birds he saw that month. Notice that Wetmore would have to be able to look at a bird and identify the species before he could record the information. A M S

```
FEBRUARY, 1904 - BIRD LIST.

    Bluejay
    Hairy Woodpecker
    English Sparrow
    White-breasted Nuthatch
    Chicadee
    Downy Woodpecker
    Crow
    Great Horned Owl
    Junco
    Tree Sparrow
    Ruffed Grouse
    Barred Owl
    Screech Owl
    Bobwhite
    American Goldfinch
    Pine Grosbeak
    Prairie Horned Lark
    Brown Creeper
    Redpoll
    Evening Grosbeak
    American Goshawk
    Acadian Owl
    Bohemian Waxwing
    American Crossbill

24
```

MAKING LISTS

Wetmore had a simple idea for his journals. He just made lists of birds he saw in a single month. In your STEAM notebook, you can make lists, too. What will you list? You could list ideas, equipment, observations, what you saw, what you heard, and much more.

struggled to keep as many people working as possible. World War II, from 1939-1945, made it hard to travel to study birds. Instead, Wetmore studied birds of the Shenandoah National Park in nearby Virginia.

After World War II ended, Wetmore could travel more to study birds. From 1946-1966, Wetmore made yearly trips to Panama. He wanted to study all the birds of Panama. Later he published *The Birds of the Republic of Panama* (1984). Wetmore wrote about 189 species and sub-species of birds that were new to science. S

During his travels, Wetmore brought back 26,058 bird and mammal skins. He collected 4,363 skeletal and anatomical specimens. Also, he brought back 201 clutches of birds' eggs. M

Many birds, animals and plants were named in his honor. There were 56 new genera, species, and subspecies of birds (both recent and fossil). Other animals named for him included mammals, amphibians, insects, and mollusks. Wetmore jokingly called these his "private zoo." S

All his life, Wetmore wrote in his notebooks. He listed birds and other animals he saw in his travels. He wrote many letters. He also used photography and left several photo albums of his work. A

Published at Age 15

Wetmore, on October 15, 1901, holds a copy of 1900 *Bird Lore* magazine. His first published article, "My Experience with a Red-headed Woodpecker," appeared in this magazine. He used the pen name of Alick Wetmore. A Read his article on pp. 148-149.

SPECIMENS: Taking Something Home

Here is Alexander Wetmore on May 31, 1953. He's looking at a table with bird skins drying. After the skins dried, Wetmore sent the specimens back to the United States. A specimen is a plant or animal that is taken back to the laboratory for more study.

You can collect specimens, too. When collecting specimens, write about when, where, how, and why you collected the specimens. Always check with officials for permission to collect specimens. S

Use this space to draw, glue worksheets, or write.

Make a list.

1.

2.

3.

4.

5.

6.

7.

8.

9.

10.

NAME: DATE:

NOTES (weather, location, etc.)

Use this space to draw, glue worksheets, or write.

NAME: DATE:

NOTES (weather, location, etc.)

Use this space to draw, glue worksheets, or write.

NAME: DATE:

NOTES (weather, location, etc.)

Use this space to draw, glue worksheets, or write.

NAME: DATE:

NOTES (weather, location, etc.)

Use this space to draw, glue worksheets, or write.

IDEAS TO TALK ABOUT

I observed

NAME: **DATE:**

NOTES (weather, location, etc.)

Martin H. Moynihan

1928-1996 ornithologist or bird scientist and behavioral evolutionary biologist
Birthplace: Chicago, IL
Education: Princeton 1950 and Ph.D 1951 Oxford University

Martin Moynihan in the United Kingdom (date unknown).

As a child, Martin Moynihan traveled in Europe. He learned to speak French, German, and Spanish. Moynihan became interested in birds at age 15. He published his first scientific paper by age 18.

Moynihan worked in Panama from 1957-1974. He helped build the Smithsonian Tropical Research Institute on Barro Colorado Island. To build the Panama Canal, some rivers were dammed to make Gatun Lake. That meant the existing tropical forest was covered with water. Only the tallest hills remained above water as islands. Barro Colorado Island is one of those islands. It's become one of the most studied tropical forests in the world. S

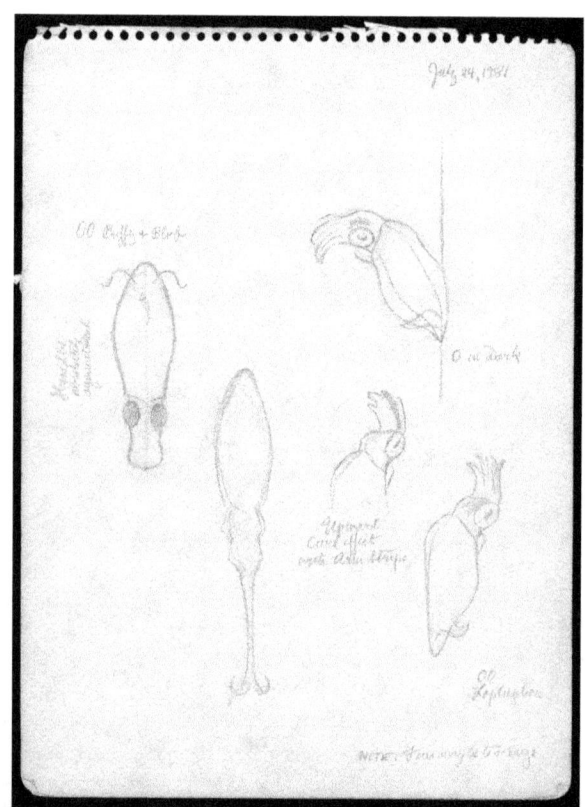

Observations of Squid in Panama

July 24, 1981. Sometimes, Moynihan wrote notes about his drawings. The first one on the left says, "Should be absolutely symmetrical." M Symmetrical means that if you folded the squid in half, both sides should look the same.

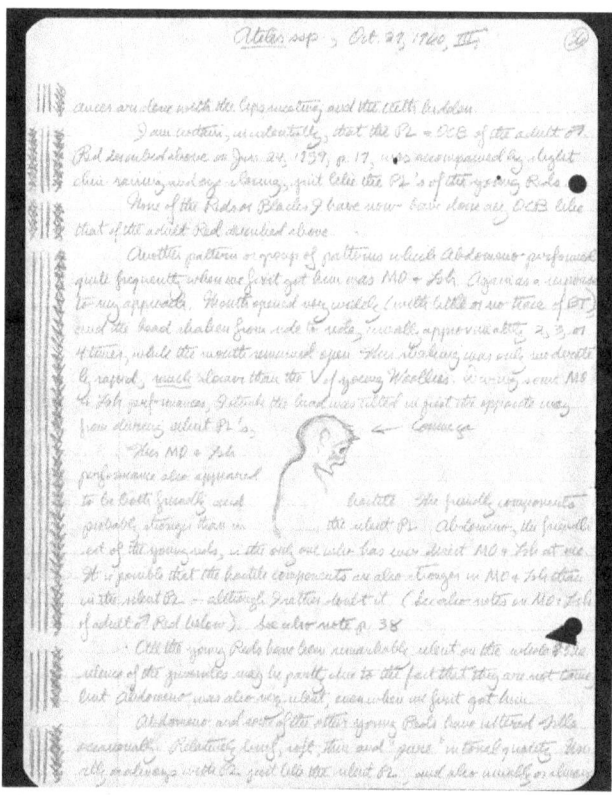

Compare Your Notes Over Time

This is Moynihan's October 29, 1960 notes on *Ateles ssp*, a primate species of Panama. Moynihan often reminded himself to compare observations over time. Here he notes a previous description on June 24, 1959 to reread later.

The Institute studied biological diversity and tropical ecology. 🅂 Biological diversity is the study of how many and what kinds of plants or animals are found in a certain area. An area with lots of different kinds of plants or animals has a high biological diversity.

Tropical ecology means the study of the ecology in the tropic zones. Tropical zones are the area between the Tropic of Cancer in the north and the Tropic of Capricorn in the south. In the tropics, scientists usually find more biological diversity.

Moynihan studied tropical biology, both on land and at sea. He often compared tropical and temperate species of animals. Temperate animals live between the tropic of Cancer and the Arctic Circle in the north. In the south, they live between the tropic of Capricorn and the Antarctic Circle.

Often scientists spend a lifetime studying just one kind of animal. But Moynihan wanted to understand how animals created social structures. How did their families work? How did their communities work? He also wondered how animals communicated, or talked, to each other. To study these things, he looked at three different kinds of animals. Moynihan studied monkeys, squid, and gulls. Then, he compared the studies from each species.

Moynihan's notebooks were always with him. He was known for black-and-white ink drawings of his subjects. 🄰 Sometimes they were published as part of his scientific papers. Underwater, he carried a waterproof notebook to take notes about squid. 🅃

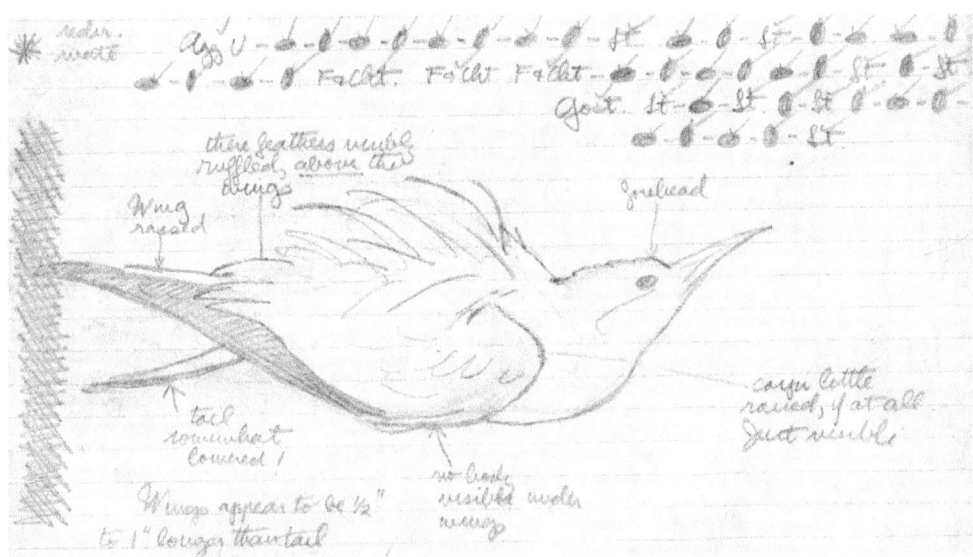

Organization of Field Notes

Moynihan carefully dated each entry in his field notes. The dates created a strong chronological order. The newest entries were added to the back of each notebook. However, every couple years, he reorganized his notebooks. He put information about a single species into a folder. Then, he renumbered the pages. This reorganization was topical. A topical order means the notes were organized by information about a certain animal. 🄰

Moynihan often drew in his notebook. Here on November 13, 1955, he labeled parts of the South American gulls (laridae). The note on the bottom left says, "Wings appear to be 1/2" to 1" longer than tail." 🄰 🅂

Some of Moynihan's notes are just words. Sometimes, he combined words and drawings. Sometimes, he just drew. He liked to use exclamation points and question marks. The next time you draw, try adding labels to help explain the drawing.

Use this space to draw, glue worksheets, or write.

Draw something.

Now, label the drawing.

Write other notes about the drawing.

NAME: DATE:

NOTES (weather, location, etc.)

Use this space to draw, glue worksheets, or write.

NAME: DATE:

NOTES (weather, location, etc.)

Use this space to draw, glue worksheets, or write.

NAME: DATE:

NOTES (weather, location, etc.)

Use this space to draw, glue worksheets, or write.

NAME: DATE:

NOTES (weather, location, etc.)

Use this space to draw, glue worksheets, or write.

IDEAS TO TALK ABOUT

Information I want to share

NAME: DATE:

NOTES (weather, location, etc.)

William Healey Dall

1845 - 1927
malacologist, or mollusk scientist, and naturalist
Birthplace: Boston, MA
Education: Studied with Harvard professors, self-taught

In 1865, Alaska was owned by Russia. It was called Russian America. Not much was known about these frozen lands. But technology changed all that. In 1858, the Western Union Telegraph Company had placed the first undersea cable across the Atlantic Ocean. By 1861 the eastern United States was linked to the west coast by telegraph. After that success, Western Union decided to start a new project. They wanted to lay telegraph cable from San Francisco, California to Moscow, Russia. **TE**

As part of the Western Union Telegraph Expedition, 20-year old William Dall traveled north. His scientific interests were mainly in malacology, or the study of

Dall dressed in the Expedition uniform, about age 20.

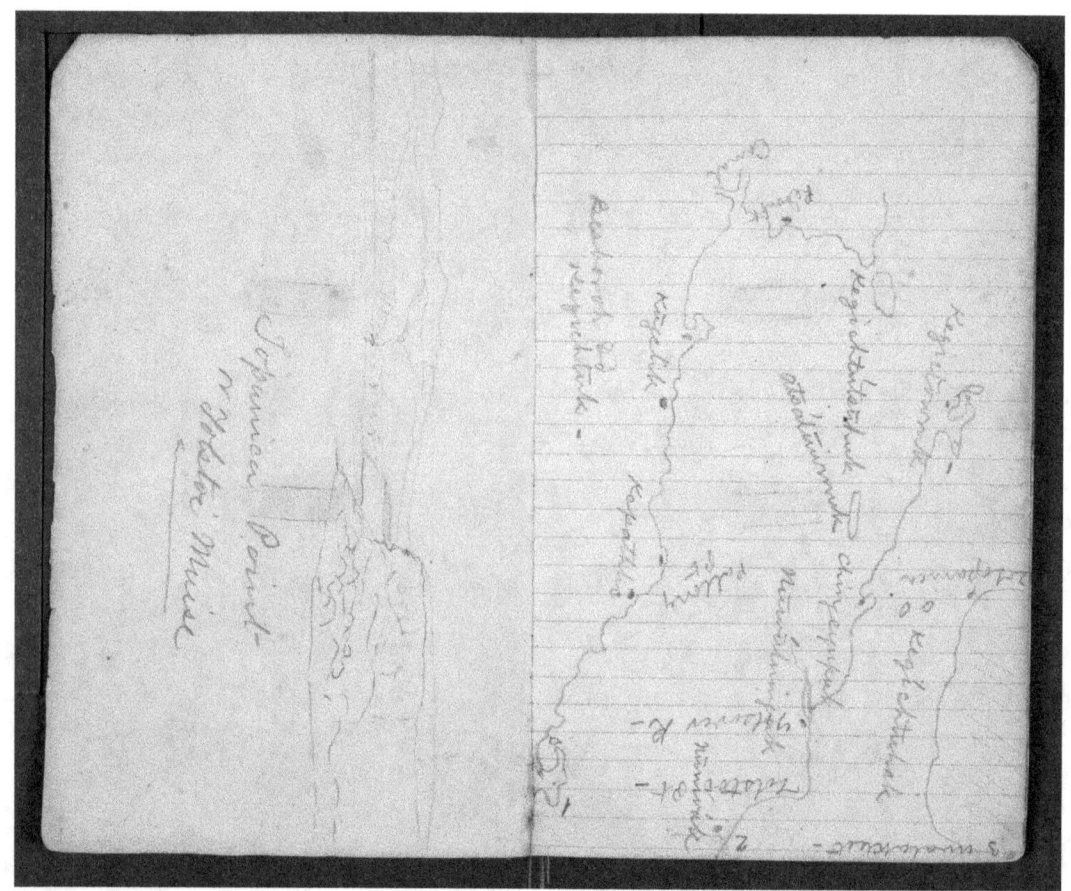

Drawing Maps

Dall drew this map of the Alaskan region where they were traveling. On the left-hand side is a sketch of a village. **A M** The names are written in Russian or native American languages. He included phonics marks to help him remember how to say the words correctly.

mollusks. The mollusk family includes snails, slugs, clams, octopus and squid. Dall kept a diary of his travels through Alaska and across the Bering Sea to Russia. While the Expedition was in Alaska, they received exciting news. On March 30, 1867, United Stated bought Alaska from Russia. They paid $7.2 million.

The Expedition's project failed. Workers couldn't install telegraph poles and telegraph lines to connect the United States and Russia. It was too hard for men to work in Alaska's cold. The frozen ground made it hard to dig a hole for a pole. First, workers had to light a fire to thaw the ground. Then, they could dig a hole. E Also, in 1867, cold weather clothing was poor compared with today. They only had wool clothing or furs like the native people.

Work on the telegraph lines stopped in 1867. The real success of the program was the scientific exploration of the region. From 1866-1868, Dall collected thousands of specimens for the Smithsonian Institution. After the United States bought Alaska, Dall took many other trips to Siberia and Alaska. He was also known for his cartography, or map making. M Dall liked Alaska so much that he and his wife honeymooned there in 1880.

Later, Dall used his notes to write an 1870 book, *Alaska and Its Resources*. In his lifetime, he described 5,427 species, many of them mollusks. Many animals were named for him including ten mollusks, the Dall's sheep and Dall's porpoise. Dall is one of the founding members of the National Geographic Society.

Ingalut pottery at Utkusik

Thursday, July 18, 1867. Can you guess how much liquid each pot holds? For the answer, see the discussion questions on page 142. A T

Native Drawing

Okanóchlŭk. A sketch of a native American whom Dall met in his travels. A

FIELD BOOKS - WRITE AND DRAW

Scientists' books are called many things: field book, experiment book, diary, travel log, lab notebook, and so on. But scientists all do the same thing. They write, diagram, or draw something about their observations and their work.

Dall drew maps of Alaska. He included rivers, mountains and other geographic features. The names are either native American or Russian.

To keep track of his travels, Dall also wrote a travel log. He started each page with a date. Next, he talked about where he was and what he was doing.

Dall drew technical diagrams of objects. Once, he stopped in a native village and wrote a description of their pottery. He drew pictures of several of the pots.

When Dall met someone new, he often drew their picture. This is a likeness of Okanóchlŭk, a native woman. Dall even wrote a poem for his sister about his travels.

When Dall came across new words, he wrote them down. Look at MY GLOSSARY starting on page 134. A glossary is a list of words with a short definition. As you make observations, write new words in your glossary. Write or look up a definition for the words.

Use this space to draw, glue worksheets, or write.

Draw something.

Use words to explain something that the drawing can't explain.

NAME: DATE:

NOTES (weather, location, etc.)

Use this space to draw, glue worksheets, or write.

NAME: DATE:

NOTES (weather, location, etc.)

Use this space to draw, glue worksheets, or write.

NAME: DATE:

NOTES (weather, location, etc.)

Use this space to draw, glue worksheets, or write.

NAME: DATE:

NOTES (weather, location, etc.)

Use this space to draw, glue worksheets, or write.

IDEAS TO TALK ABOUT

I expected

Instead

NAME: DATE:

NOTES (weather, location, etc.)

Joseph Nelson Rose
1862-1928 botanist or plant scientist
Birthplace: Liberty, Indiana
Education: Ph.D. 1889, Wabash University

Joseph Nelson Rose at age 23

Joseph Rose spent a lifetime describing plants. While at Wabash University, Rose learned to classify plants. He helped classified such herbs and spices as dill, caraway, fennel, and coriander. S

In 1883, he moved his wife and six children to Washington, D.C. There, he worked as assistant botanist at the United States Department of Agriculture. He began studying plants of Mexico and Central America. He specialized in cacti. He visited Mexico nine times on collection trips and sent back many live specimens, which were grown in greenhouses. Some of the cacti he collected are still alive. For more, see http://huntington-blogs.org/2013/07/a-landscape-by-the-numbers/

Rose co-authored a four-volume book, *The Cactaceae*,

```
05.363     Echinocactus myriostigma S.-D.
     Flowered April 30, 1912; flower 3.5 cm. long, about 1.5 cm. in diame-
ter; petals numerous, oblong elliptical, jagged, glistening creamy white
on inside, same on outside save for narrow purple stripe at center; bracts
acuminate, pinkish to dark purple, bearing wool; stamens numerous, outer
ones attached higher up throat and standing higher; filaments white, weak;
anthers golden yellow with abundant pollen; pistil longer than stamens,
separated into 7 green stigma lobes, about 6 mm. long, puberulent; ovary
globular, circular, 2 to 3 mm. deep; no tube proper; throat funnel shaped,
1.5 cm. long and 1 cm. wide at top.
     Plant collected in northern Mexico by C. A. Purpus in 1905.
                                              Fitch.
```

Laboratory or Greenhouse Observations

Not all observations are done in the wild. This is a careful observation of a blooming cactus in the greenhouse. Many of the words or vocabulary used are scientific ways of describing a plant. For example, a bract is a leaf that has a special purpose. Notice that Rose describes:

Size: flower 3.5 cm long, about 1.5 cm in diameter
Shape: petals numerous, oblong elliptical, jagged
Color, inside and out: glistening creamy white on inside, same on outside save for narrow purple stripe at center
Number: separated into 7 green stigma lobes
Place collected: Northern New Mexico by C.A. Purpus in 1905

with Lord N. L. Britton. The watercolor and line illustrations were done by Miss Mary Eaton, who worked for $300/year. To write the book, Rose and Britton traveled across Europe to visit herbariums. An herbarium is any collection of dried plants that are arranged in some order. They also traveled to countries in South America: West Indies, Chile, Peru, Bolivia, Argentina, Brazil, Venezuela, and Ecuador.

Rose helped name and describe 972 species of cacti. To do this, he looked at all the similar species. Next, he described the new species carefully. He made sure the description was different from other similar cacti. Before Rose and Britton's book, many cacti were classified in confusing ways. Their book made classification easier to understand. The cactus book encouraged other scientists to be interested in the proper scientific classification of plants.

Rose also worked as curator at the National Herbarium. The Herbarium is part of the Smithsonian Museum of Natural History. The National Herbarium sponsored expeditions to various places to collect plants. It currently holds over 5 million specimens. About 4.5 million of the specimens are plants pressed onto paper. The plants come from all over the world, including the neotropics, North America, Pacific Oceanic islands, the Phillipines and the Indian subcontinent.

In Rose's lifetime, he published over 200 scientific articles. He co-wrote with at least twelve other botanists.

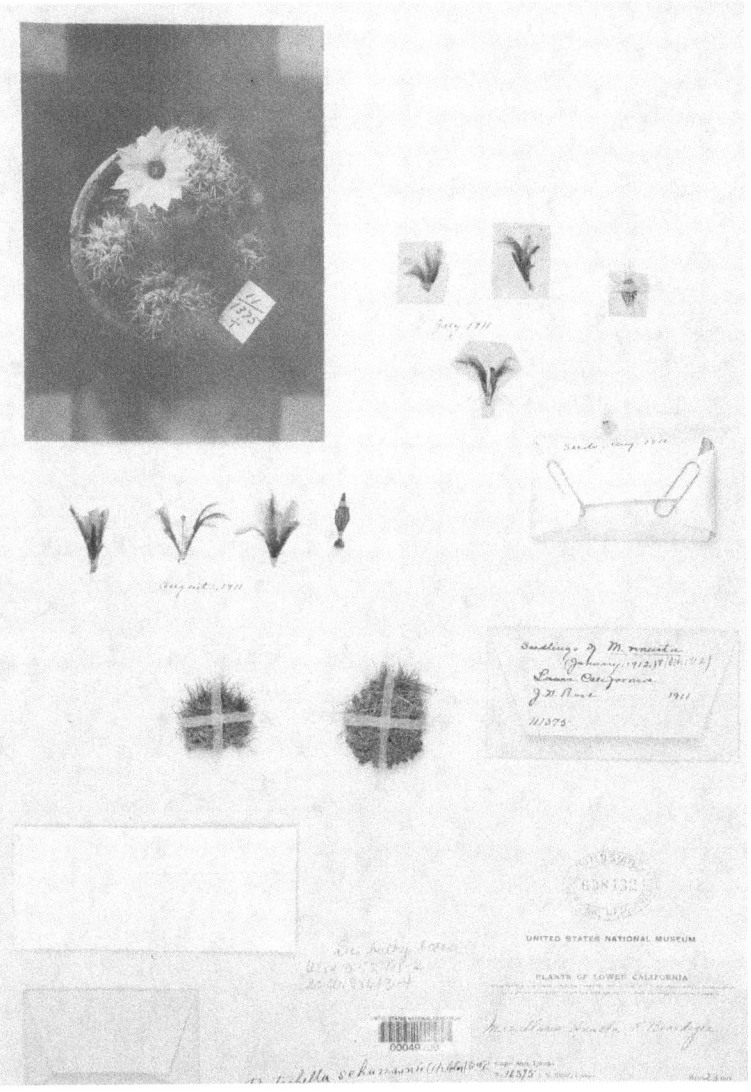

Drying Plants

Most plant specimens are preserved by drying. Plant presses are built to squeeze a plant until the water is taken out. Dried plants will not rot as easily. Plants are placed between absorbent papers. Usually a press has screws to squeeze the plant tighter and tighter. Scientists might dry the whole plant. Or they might dry a plant part, like a flower. As it dries, the papers are changed often. When the paper is changed, a scientist might rearrange the plant so the interesting parts are showing. For example, a flower might be turned so the petals press flat.

After a plant is dried, it's mounted on heavy paper. The goal is to show as much of the plant as possible for future studies. Notice that the photo shows envelopes holding seeds.

Use this space to draw, glue worksheets, or write.

Describe something.

Size

Shape

Color

Number

Where observed

Did you use scientific vocabulary? If so, define the words.

NAME: **DATE:**

NOTES (weather, location, etc.)

Use this space to draw, glue worksheets, or write.

NAME: DATE:

NOTES (weather, location, etc.)

Use this space to draw, glue worksheets, or write.

NAME: **DATE:**

NOTES (weather, location, etc.)

Use this space to draw, glue worksheets, or write.

NAME: DATE:

NOTES (weather, location, etc.)

Use this space to draw, glue worksheets, or write.

IDEAS TO TALK ABOUT

I think

My reasons are

NAME: DATE:

NOTES (weather, location, etc.)

Lucile Mann
(1897-1986) editor
Birthplace: Ann Arbor, Michigan
Education: B.A. 1918 University of Michigan

William (Bill) Mann
(1886-1960) entomologist or insect scientist, zoologist, and zookeeper
Birthplace: Helena, Montana
Education: B.A. 1911 Stanford University, Sc.D 1915 Harvard

As a child, William Mann loved animals. Once, he tried to run away and join a circus. At 17-years-old, he worked briefly at the National Zoological Park in Washington, D.C. He cleaned animal cages. After college, he worked as an etymologist from 1916-1925. In 1925, he became the Superintendent of the National Zoological Park (NZP). He led many expeditions to collect live animals for the zoo.

Manns, 1931 on expedition to British Guiana.

Life at the Zoo

As the wife of a zookeeper, Lucile Mann became familiar with orphaned or sickly animal babies. When needed, the Manns brought home baby animals to feed them and take care of them. This cub is named Babette.

For many years Lucile edited the *Tiger Talk*, the NZP's newsletter. She also edited *Spots and Stripes*, the Friends of the National Zoo newsletter.

For most of the overseas live-animal collection trips, Bill and Lucile traveled together. While Lucile didn't have a degree in biology, she learned on their trips. She wrote three books: *Tropical Fish: A Practical Guide for Beginners, From Jungle to Zoo: Adventures of a Naturalist's Wife,* and *Friendly Animals: a Book of Unusual Pets.*

"Have a chocolate, anyway." Oral Histories

How did Bill and Lucile Mann meet? Listen to audio recordings of Lucille telling that story here: http://siarchives.si.edu/blog/life-wild-side-lucile-quarry-mann

Bill and Lucile Mann made a great team in their works with animals. Together, they traveled, raised baby animals, worked at the NZP, and enjoyed animals. The Manns loved to go to the circus. Lucile liked riding the elephants best. Besides raising baby animals, Lucile also took care of Bill's live snake collection at their apartment.

Lucile Quarry worked during World War I for Military Intelligence in Washington, D.C. By 1922, she was an editor at *The Woman's Home Companion*. Lucille and William married in 1926. Lucile often traveled with William on expeditions to collect live animals. She helped raise some of the baby animals born at the NZP. Once, she had to hide a bag of live snakes under her skirts while on a train trip. She was a member of the Society of Women Geographers.

One big trip to collect animals was the 1937 National Geographic Society-Smithsonian Institution Expedition to the Dutch East Indies. The trip lasted from January to September. First, the expedition rode a train ride across the U.S. They sailed from Seattle on January 19. The Manns stopped in Japan, where the newspapers reported daily on their sightseeing stops. They had tea with the Japanese prince at the Imperial Palace. They visited many Japanese zoos. After Japan, the Manns stopped in Hong Kong and Singapore. They arrived in Sumatra on March 2nd. Their stay in the Dutch East Indies (now Indonesia) included visits to British India, Malay, and Siam (now Thailand). After over eight months of travel, the expedition returned to Washington, D.C.

Collectors brought back 879 live specimens. There were 169 species of animals, birds and reptiles. Some of the animals didn't survive the journey. But the expedition added many Asian species to the NZP collection.

Diary: Experiencing New Cultures

Lucile Mann kept a typed diary of their 1937 trip to Sumatra. In Singapore, Lucile wrote about the city. A syce is a servant. A zebu is a hump-backed cow used to pull carts.

```
     As we drove back to the hotel, I tried to count the smells
of Singapore:  Incense, fried fish, wood smoke, the oil on the
syce's hair, roasting peanuts, the scent of flowers, the smoke of
firecrackers which the Chinese are always putting off, and occasion-
al unsanitary whiffs better not analysed.

     We were sorry to leave this fascinating city, with its mixture
of races, its crowded harbor, and the waterways where so many
people live their lives in sampans;  traffic policemen with rattan
boards on their back for stop and go signs; sikhs and tamils from
India directing the traffic of every imaginable Asiatic people;
zebu carts rubbing axles with the latest make of motor car;  orchids
a customary decoration on the table.
```

When you observe things for your STEAM notebook, pay attention to your senses. Write about what you see, hear, smell, touch (temperature and texture, not feelings), and taste. Always be sure something is safe to taste or touch.

Lucile listed eight things she smelled in Singapore. Think about a specific place and write a list of eight things you would smell.

The second paragraph is a list of things Lucile saw in Singapore. She included many specific details. For example, she didn't just write "traffic policemen." She added details about the signs they carried.

Think about the specific place you chose for smells. Write eight things you would see there. Can you also write eight things you hear? Choose only the best sensory details and write a paragraph about the place.

Use this space to draw, glue worksheets, or write.

Use Your Senses.

I see

I hear

I smell

I feel (temperature and texture, not feelings)

I taste (only taste things you know are safe)

NAME: DATE:

NOTES (weather, location, etc.)

Use this space to draw, glue worksheets, or write.

NAME: DATE:

NOTES (weather, location, etc.)

Use this space to draw, glue worksheets, or write.

NAME: DATE:

NOTES (weather, location, etc.)

Use this space to draw, glue worksheets, or write.

NAME: DATE:

NOTES (weather, location, etc.)

Use this space to draw, glue worksheets, or write.

IDEAS TO TALK ABOUT

I found

My evidence is

NAME: **DATE:**

NOTES (weather, location, etc.)

FRED SOPER 1893-1977

epidemiologist or scientist who studies the source and cause of infectious diseases
Birthplace: Kansas
Education: B.A., M.S. University of Kansas, M.D. 1918 Rush Medical College/University of Chicago

Fred Soper in 1928.

Fred Soper was a public health doctor. After his medical training, Soper started work at the Rockefeller Foundation in January, 1920. Soper started his career in public health at a time when major diseases spread across the globe including hookworm, tuberculosis, malaria, influenza, typhus fever, and yellow fever.

Among the Rockefeller projects was the International Health Division (IHD). They had worked to eradicate, or completely destroy, hookworms in the American south from 1909 to 1914. In 1910, over 40% of people had hookworms, a worm that can live and grow in human intestines. M The program was successful. The IHD

Public Health Issues: Hookworms and Tapeworms

One of Fred Soper's first public health issues was to rid Brazil of hookworms and tape worms. First, they treated patients to get rid of the worms. This 9-year-old Brazilian boy is holding a board with the worms that he expelled. But Soper didn't just treat people. He also worked to make sure they didn't get sick again. That is called preventative medicine. S After treating hookworm and tape worms in people, Soper worked with cities, villages, and communities to build latrines. With clean bathrooms, worm infections went down.

decided to repeat the program in other countries.

Soper worked in Brazil from 1920-1927 to control hookworm infections. In 1927, Soper worked to stop malaria and yellow fever. Many diseases are carried by mosquitoes. Some people called Soper the "Mosquito Killer." One tool to kill mosquitoes was a new man-made chemical called DDT. 𝕋 Many called DDT a lifesaver because it killed so many mosquitoes. However, DDT had unexpected environmental effects. DDT was banned in 1972. Today, we still have many diseases carried by mosquitoes.

During World War II, Soper worked with military health organizations. Soper led the fight against typhus in the Middle East and North Africa. Typhus wasn't carried by mosquitoes. Instead, it was carried by head lice. 𝕊 Again, DDT was used to kill the insects. He also worked to kill mosquitoes in Egypt and Italy. After the war, Soper returned to South and Central America. There, he worked on malaria treatment and smallpox vaccinations.

Throughout his life, Soper worked on many public health issues: sanitation, vaccination, diagnosis and treatment, maternal and child health, and health education. 𝕊 Today, public health faces many of the same problems. Hookworms are rare in the United States. Worldwide, about half a billion people are infected with hookworms. Many diseases are still carried by mosquitoes, such as the Zika virus. Head lice are common world wide. The public fight for healthy living conditions continues.

> *A New Idea.* Arriving in Sertão, hot tired and dusty, and inquiring for the latrine, I was proudly escorted by the hoteleiro, not to the kitchen as you might have expected, but la' fôra no campo, to a solid structure built to withstand the wear and tear of generations. It is true that it lacked a pit and was muito defeituosa being open to flies, chickens and pigs. However it is provided with two departments and is a beginning. I believe it is the only latrine in the villa. Being since early youth intensely interested in that branch of American literature, which may be safely designated here as 'Latrine Poetry,' I have let no opportunity escape me to become familiar with the faint beginnings of a similar literature in Brasil. Noting that something had been written on the wall, I commenced to read and encountered not the efforts of a budding poet, but a new idea. This is what I read.
>
> Setembro de 1921, foi solemnemente inaugurada esse indispensavel melhoramento da villa do Sertão, com a presença das pessoas assignantes: "   " and below were the names of twenty one men among them one engineer and others of note. Names of senhoras were not found in the list. Investigating the matter further it was found that the latrine had been constructed as a result of propaganda of the C.R. and that the inauguration had been scheduled by a viagante who had visited the post in Torres. We were assured that the inauguration was a really festive occasion with cerveja, muito uso da palavra and other requisites for a successful festa. We assured the intendente with all gravity that we were heartily for the system of official and public inagurations but that we thought that they should be made more liberal and that the ladies should be invited as we considered the installation of latrines in this district as of more public importance than the installation of churches schools bridges roads and other melhoramentos. In fact I waxed so eloquent that in a moment of forgetfulness I offered to come from Porto Alegre (a small matter of four day's trip) com a senhora to assistir the next inaururation provided there were ten or more new latrines to be installed.

A New Idea

In 1922, Soper visited the city of Sertão, Brazil. There he learned a startling thing. Recently, the city had a big party. They were celebrating the community's first public latrine. A latrine is a community bathroom. Usually, they are made by digging a hole in the ground.

Latrines were an important part of fighting hookworms. Hookworms are worms that live and grow in the intestines of people. When an infected person goes to the bathroom, the hookworm's eggs come out, too. The eggs hatch and become larvae. The larvae can enter people through the bottoms of their feet. People walking barefoot in unclean areas can easily be infected with hookworms. Latrines meant the hookworm eggs weren't spread in the soil.

When Sertão's first public latrine was opened, they held a party. There was food and many speeches. In his diary, Soper joked that a celebration party was a "new idea." He made a promise. If the region opened at least ten new latrines, Soper would make the four-day journey to join the next party. Sometimes scientists wrote funny things in their notebooks!

Use this space to draw, glue worksheets, or write.

When you write, use these ideas.
Use sensory details.

Use interesting words.

Vary your sentence length.

Use humor or interesting ways to say things.

NAME: **DATE:**

NOTES (weather, location, etc.)

Use this space to draw, glue worksheets, or write.

NAME: DATE:

NOTES (weather, location, etc.)

Use this space to draw, glue worksheets, or write.

NAME: **DATE:**

NOTES (weather, location, etc.)

Use this space to draw, glue worksheets, or write.

NAME: DATE:

NOTES (weather, location, etc.)

Use this space to draw, glue worksheets, or write.

IDEAS TO TALK ABOUT

I noticed

This is an example of

NAME: DATE:

NOTES (weather, location, etc.)

Mary Agnes Chase

1869-1963 agrostologist, or a plant scientist specializing in grasses, suffragist or someone working for women's rights, and author
Birthplace: Wady Petra, Illinois
Education: Self-taught

Mary Agnes Chase and Brazilian botanist Dona Maria Bandeira at the top of Mt. Itatiaia in 1924. Chase set records for a woman climbing the highest mountains in Brazil.

Mary Agnes Chase worked for sixty years as a botanist. She specialized in grasses. Her husband, William Ingraham Chase, died in 1888 of tuberculosis. They had only been married for one year. She never remarried.

In 1893, Chase visited the Columbian Exposition in Chicago, Illinois. It inspired her to study plants and grasses. She worked at the Chicago stockyards as a meat inspectress. For her work, she learned to use a microscope. T She also used the microscope to look at plants. It helped see plants in detail. She became a botanical illustrator, drawing plants and animals.

In 1903, Chase started work for the U.S. Department

Botanical Illustrator, Expert, Author

Chase began her career as a botanical illustrator in Chicago. When she moved to Washington, D.C. in 1903, her job was drawing more illustrations. The careful study needed to draw plants, though, meant she learned a lot about them. Even more study meant she became an expert about grasses.

As an expert, she could also write about grasses. She moved from artist to expert to author. S A

Her first book was published in 1922, *The First Book of Grasses, the Structure of Grasses Explained for Beginners*. For the book, she drew her own black-and-white illustrations. The illustrations helped to explain details about grasses. This page includes four drawings. They compares the grasses and show contrasts, or differences.

She wrote in the book: "Of all plants, grasses are by far the most important to man. The grains of wheat, barley, rye, oats, rice, corn sorghum, and millet form the staple food of the greater part of mankind...The grains are also the sources of starch, alcohol, and glucose."

of Agriculture (USDA) as a botanical illustrator. She soon moved up to a botanist position. Because of her work, Chase published scientific papers about grasses. Between 1905 and 1912, Chase studied and collected grasses of the United States. She collected grasses along the United States eastern and southern coasts and in the southwestern states. That research led to the 1922 book, T*he First Book of Grasses, the Structure of Grasses Explained for Beginners.* S A

In 1911 and 1912, Chase was not allowed to travel with a research group to Panama because she was a woman. For many years, she worked with women's rights groups known as suffragists. She was jailed several times for her protests.

Besides United States grasses, Chase also studied grasses in Europe, northern Mexico, Puerto Rico, Venezuela and Brazil. In 1924-1925, she traveled for eight months in Brazil. She collected over 500 new species of grasses. She brought back more than 19,000 other specimens. Usually, Chase pressed the plants, drying them to preserve them. T E Each specimen was carefully cataloged for the species, location collected and other important facts. She returned to Brazil in 1929-1930 and collected even more grasses. She retired in 1939 from the USDA. After that, Chase served as the Honorary Curator of the United States National Herbarium at the Smithsonian Institution.

In 1958, at the age of 89, she was awarded an honorary doctorate granted by the University of Illinois. It was her first and only degree.

Snakes and Adventures

In this 1929 journal, Chase writes a narrative A of her travels. She climbed the western side of the Caparaó Mountain in Brazil. She hoped to find new grasses. She joked, "One who has been to Brazil is expected to tell about snakes, so I am glad to have at least this big one." About eight feet away, she saw a 15-foot snake. Its head was raised two feet off the ground. She wrote, "I backed off respectfully."

Mrs. Mexia of the University of California climbed the mountain with her. It rained constantly. At one point, Mrs. Mexia became too tired to keep going. The guides pitched a tent and tied it down. Then the guides left. The two women were on their own for two nights of storms. They refused to leave early because they were still collecting and pressing grasses. Chase wrote, "...the plants were what we came for..."

After the second stormy night, the guides returned and escorted the women to safety at a herder's house.

Use this space to draw, glue worksheets, or write.

Draw an outline of an object.
Add details to the outline.
Use a magnifying glass or microscope to add more details.

NAME: **DATE:**

NOTES (weather, location, etc.)

Use this space to draw, glue worksheets, or write.

NAME: DATE:

NOTES (weather, location, etc.)

Use this space to draw, glue worksheets, or write.

NAME: **DATE:**

NOTES (weather, location, etc.)

Use this space to draw, glue worksheets, or write.

NAME: DATE:

NOTES (weather, location, etc.)

Use this space to draw, glue worksheets, or write.

IDEAS TO TALK ABOUT

I observed

This happened because

NAME: DATE:

NOTES (weather, location, etc.)

Donald S. Erdman [1]

1919 - 2002 ichthyologist or fish scientist
Birthplace: New Jersey
Education: unknown

Assumed to be Donald S. Erdman, 1949 in Saudi Arabia.

Little is known about Donald Erdman's personal life. He was a conscientious objector during World War II. As a member of the Friends church, he believed in peace and not fighting. He would not become a soldier. Instead, the government let such men work in the Civilian Public Service program. First, he worked in Massachusetts and New York. There, he did forestry work, cleaned up after a hurricane, and fought fires. In New Hampshire, he worked in a mental hospital. Finally, in Puerto Rico, he did other public health work. This was likely Erdman's first time in Puerto Rico. He later returned there to live and work.

After World War II, Erdman served as a Scientific Aid in the Division of Fishes, United States National Museum

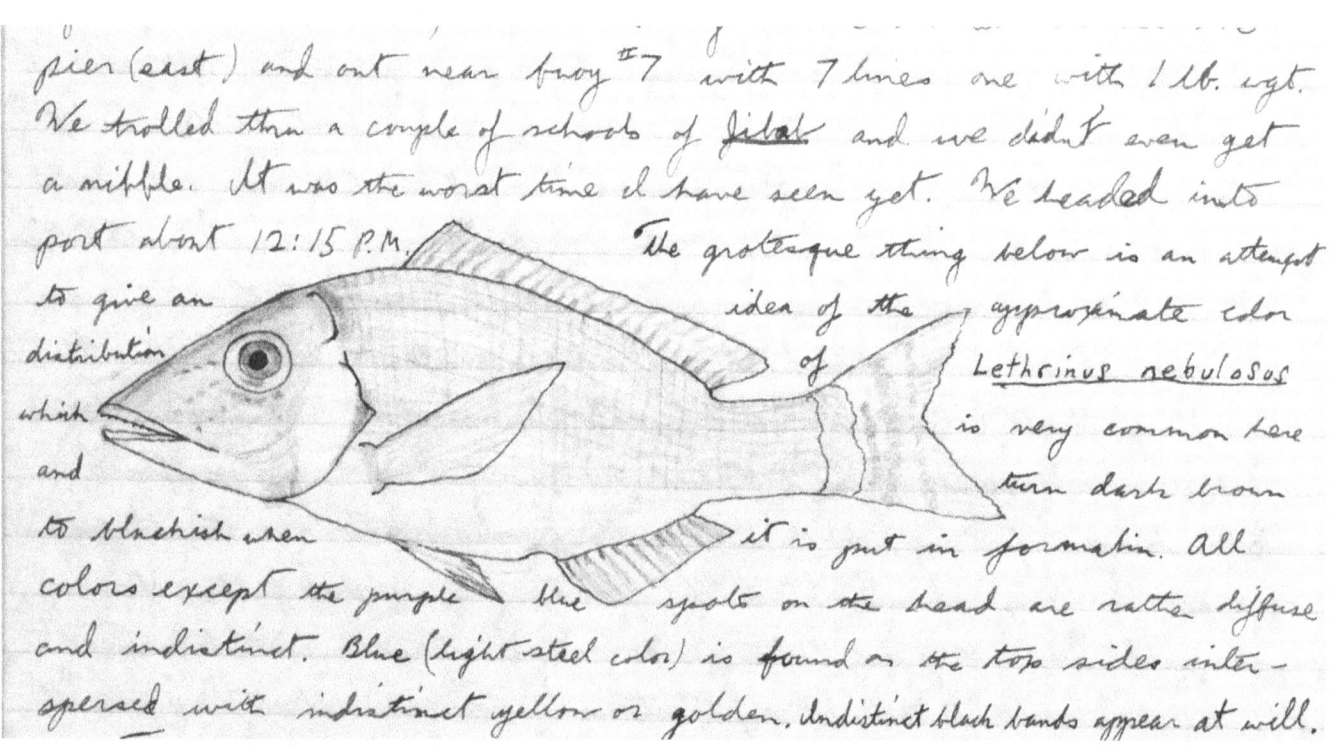

Color Drawings

This drawing was originally in color. Shown here in black-and-white, it's a drawing of a Spangled Emperor fish (or Green Grouper - *Lethrinus nebulous*). Erdman drew in color because of a problem with preserving the fish. When the fish were preserved, they turned dark brown or black. Erdman explained that the colors were spread out except the purple blue spots on the head. The light steel-blue color was found on the top, mixed in with yellow. Faint black bands seemed to be random. 𝔸 𝕊

(USNM). In 1948, Erdman joined a fisheries survey of the Persian Gulf and Red Sea. They were invited by the Arabian American Oil Company (ARAMCO). ARAMCO wanted to know more about local fish. Perhaps, they could buy local fish to feed their employees. Erdman collected nearly 5,000 fish specimens for the USNM. For most, he wrote down the English name and the Arabic name. After the survey, ARAMCO decided to buy more fish from local Arab fishermen.

In the mid-1950s, Erdman returned to Puerto Rico. He studied fish and marine mammals such as blue marlins and whales. He wrote many scientific papers about his studies. His 1972 book, *Inland Game Fishes of Puerto Rico*, is still a classic today. His last known publication was in 1986 for the *Tropical Fish Hobbyist* magazine.

During his trip to the Persian Gulf for ARAMCO, Erdman kept a journal. It included notes, drawings, photos, and articles. Erdman wrote, "The Persian Gulf and Red Seas can in summer be the hottest seas in the world."

Vast sun-baked deserts surrounded the area. The deserts helped heat up the water. In August 1948, Erdman recorded an open water temperature of 39C° (approx. 104F°). He was near Chaschuse Island in the Persian Gulf. The island was about two miles from the mainland of Saudi Arabia. The water was so hot that it was uncomfortable to wade. In spite of the heat, Erdman saw many small fish. Even ponds on the island had many fish.

Trapping Fish

Erdman often watched Arab fishermen at work. They used a variety of ways to fish. Sometimes, they used pot traps on the sea floor. When they used a hook and line, they didn't use a pole. At night, they used a 100-watt bulb to attract fish.

The Arab fishermen also used traps that take advantage of the high tides. The strong stems of large date palm leaves were shredded of their leaflets. The ends were stuck in the muck sand. One stem was tied closely to the next, until a whole trap was formed. The traps worked because fish swim toward shore with the high tide. When the tide ebbs or goes out, the fish go out, too. At high tide the tops of the palm stakes are almost under water. At low tide, you can easily see the stakes.

Erdman explained, "In Tarut Bay the traps are arranged in a wide 'V', the open part toward the shore. At the angle of the V is a rounded trap box where the fish are recovered at low tide." He said that to catch the fish, the fishermen waded into the trap. They used nets to catch the fish.

Use this space to draw, glue worksheets, or write.

Draw an outline of an object.
Use color to add details.
Use color to add information to the drawing.

NAME: DATE:

NOTES (weather, location, etc.)

Use this space to draw, glue worksheets, or write.

NAME: DATE:

NOTES (weather, location, etc.)

Use this space to draw, glue worksheets, or write.

NAME: **DATE:**

NOTES (weather, location, etc.)

Use this space to draw, glue worksheets, or write.

NAME: DATE:

NOTES (weather, location, etc.)

Use this space to draw, glue worksheets, or write.

NAME: DATE:

NOTES (weather, location, etc.)

Robert E. "Bob" Silberglied

1946-1982 etymologist or insect scientist
Birthplace: Brooklyn, NY
Education: B.S 1967, M.S. Cornell University, Ph.D. 1973 Harvard University

Bob Silberglied, March 20, 1970. Sitting on a Galapagos Tortoise.

Bob Silberglied was always interested in entomology. At 12-years-old, he made a precise pen-and-ink drawing of a tiger swallowtail butterfly. At 15-years old, his first scientific paper was published. It was about *Drosophila melanogaster*, the common fruit fly.

At 19-years-old, Silberglied joined an "Ant Safari" with Cornell University. It was a 41-day expedition to study insects. His notebook records that they left Ithaca, NY at 10 a.m. on July 10, 1965. The odometer read 16027. They arrived home on August 20, 1965 at 2:30 p.m. His last recorded mileage was in Greenfield, Indiana at 22643

Photography, Optical Microscopy, Ultraviolet Photography S T E A

Ultraviolet photography captures ultraviolet light, or UV light. UV light has short wave lengths. Humans can't see UV light, but special cameras can. Sometimes, cameras and microscopes are combined. This lets a photographer take pictures of very small things. Here, Silberglied photographed a close up of a butterfly's wing. It shows the overlapping scales.

miles. Can you calculate how many miles he traveled? M

On the Ant Safari, the Cornell team planned to use black lights, or ultraviolet lights (UV-A). They hoped the lights would attract insects. T Silberglied kept careful notes about when and where they set up black lights. Each day, he recorded their catch.

Silberglied worked as a professor at Harvard University and at the Harvard Museum of Comparative Zoology. Later, he worked for the Smithsonian Tropical Research Institute (STRI) in Panama. He studied butterflies in Panama.

In his work, Silberglied used many kinds of photography. T The Ant Safari used black lights or ultraviolet lights. Silberglied learned about ultraviolet photography. Butterflies see in the ultraviolet range, but humans do not. Ultraviolet patterns are invisible to the human eye. By photographing in ultraviolet, he could see what butterflies see. Ultraviolet patterns help butterflies communicate, find nectar, camouflage, and find a mate.

Silberglied also worked with cinematography, or filmmaking. He used cinematography to slow down a butterfly's flight. Another technology that interested Silberglied was macro-photography. That means taking extreme-close-up photos of something. Macrophotography let Silberglied study the structure of a butterfly's wings.

Sadly, Silberglied died in an airplane crash in Washington, D.C. on January 13, 1982. He was only 35-years-old. The *Panamamyia silbergliedi* butterfly was named in his honor.

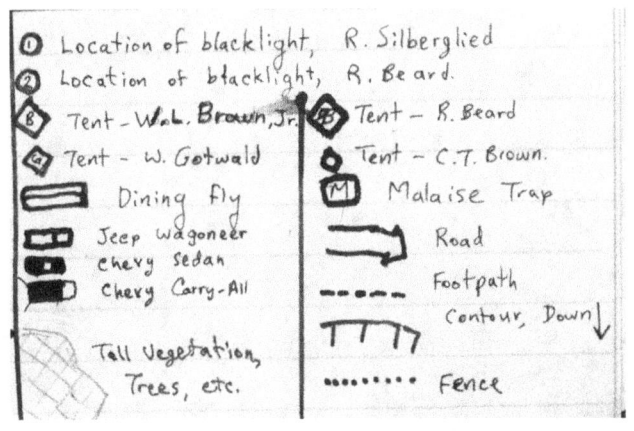

Map Key, close-up.

Outline Drawings and Map Keys A E M

In his notebooks, Silberglied included sketches, photos, diagrams, and maps. This map shows the black lighting location for July 20, 1965. The Ant Safari was at Gomez Farias, Tamps, Mexico. It was about 12 kilometers west of Mexico Highway 85. Silberglied set up at Location 1 (near center). Another scientist, Bob Beard, set up at Location 2 (to the left). Notice the compass on the map that points to the north.

Silberglied's map includes a key. A map key explains the symbols and colors on the map. Here are some things on the key:

circle - location of black light
crosshatching - tall vegetation, trees, etc.
wide arrow - road
dashed line - footpath

Draw a map. Decide what information is important and create a key for your next map.

Use this space to draw, glue worksheets, or write.

Draw a map.
Add a compass to show North.
Add a key to show something on the map.

NAME: DATE:

NOTES (weather, location, etc.)

Use this space to draw, glue worksheets, or write.

NAME: DATE:

NOTES (weather, location, etc.)

Use this space to draw, glue worksheets, or write.

NAME: DATE:

NOTES (weather, location, etc.)

Use this space to draw, glue worksheets, or write.

NAME: DATE:

NOTES (weather, location, etc.)

Use this space to draw, glue worksheets, or write.

NAME: **DATE:**

NOTES (weather, location, etc.)

Watson M. Perrygo

1906-1984, taxidermist, field collector, and exhibits specialist
Birthplace: Washington, D.C.
Education: self-taught

Watson M. Perrygo preparing a snake for a museum exhibit.

Watson M. Perrygo became interested in natural history as a teenager. He started visiting the United States National Museum (USNM). At the time, Alexander Wetmore was the Director of the USNM. While in high school, Perrygo went on birding trips with Wetmore.

As a teenager, Perrygo also saw an advertisement on a matchbox for a special class. It was for a correspondence course to learn taxidermy. Taxidermy is the art of preparing, stuffing and mounting animal skins to make them look life-like. Taxidermy was an important part of collecting specimens. Plants and animals will rot unless something preserves them. Today, preserved specimens are the only way to study some extinct species.

Preparing Specimens in the Field

Here, Perrygo is preparing specimen in Don Don, Haiti in 1928. The Parish-Smithsonian Expedition camped in tents while collecting specimens. It helped to have a taxidermist in camp. He could start the process of preserving specimens right away. Perrygo's trip notebooks include practical lists such as camp equipment, expense accounts, and travel permits.

Perrygo started working at the USNM in 1927 as a scientific aide. Later, he worked as a taxidermist. He never worked anywhere else.

Part of Perrygo's job was to collect specimens. He collected specimens in United States in West Virginia, Tennessee, Kentucky, North Carolina, South Carolina, and Arizona. He also went twice to Haiti with the Parish-Smithsonian Expeditions. Those expeditions collected 558 bird specimens. They collected some iguanas. From 1946-53, Watson also went yearly with Wetmore to Panama. In Panama, he also did some documentary videos.

Perrygo worked on many famous specimens in the USNM. Here's one example. Passenger pigeons are an extinct bird species. The last known passenger pigeon died in 1914. She was named Martha. Perrygo worked on exhibits in Bird Hall that included Martha.

During the 1950s Perrygo worked with the USNM Exhibits Modernization Program. Many new exhibits showed an animal he had preserved. They worked to show animals in a life-like way.

Perrygo and others at the USNM wrote information labels about each exhibit. Sometimes, they read scientist's notebooks or talked to the scientists about the specimen. Just like Dall's pots, the exhibits often needed both the displayed specimen and text to explain it.

Write an informational text or essay. Re-read *My STEAM Notebook*. Look for information to include in your writing.

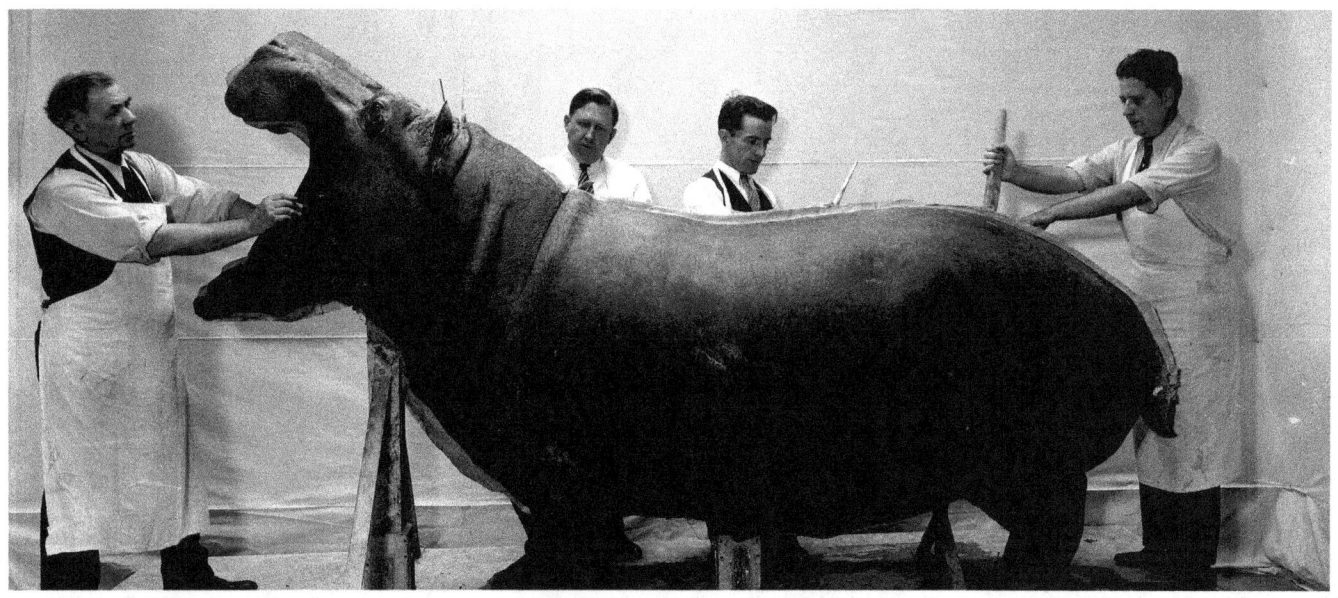

Informative Writing

This photo shows taxidermists (left to right) Julian S. Warmbath, Charles R. Aschemeier, Watson M. Perrygo, and William L. Brown mounting a hippopotamus for exhibition in the 1930s. Because the hippopotamus is a water animal, it was especially hard to prepare the skin.

Here's the informational text that accompanied the Hippopotamus exhibit.

> The "hippo" or "river horse" formerly occurred in all prominent rivers, streams, pools and lakes throughout Africa south of the Sahara. Today this animal is found only in larger bodies of water. Living in herds of from a few to many individuals, hippos spend their days lying partly or wholly submerged in water while at night they go on along well-established pathways to graze. Although the hippo looks harmless, old males may be violent. They have been known to attack canoes and small boats and smash them with their enormous jaws.

Use this space to draw, glue worksheets, or write.

NAME: **DATE:**

NOTES (weather, location, etc.)

Use this space to draw, glue worksheets, or write.

NAME: DATE:

NOTES (weather, location, etc.)

Use this space to draw, glue worksheets, or write.

NAME: **DATE:**

NOTES (weather, location, etc.)

Use this space to draw, glue worksheets, or write.

NAME: DATE:

NOTES (weather, location, etc.)

Use this space to draw, glue worksheets, or write.

NAME: **DATE:**

NOTES (weather, location, etc.)

My Glossary

Word: science notebook
Definition: a book where scientists record observations, ideas, questions, lists of equipment, lists of specimens or other information about their work

Word:
Definition:

Word:
Definition:

Word:
Definition:

Word:
Definition:

Word:
Definition:

My Glossary

Word:
Definition:

Word:
Definition:

Word:
Definition:

Word:
Definition:

Word:
Definition:

Word:
Definition:

My Glossary

Word:
Definition:

Word:
Definition:

Word:
Definition:

Word:
Definition:

Word:
Definition:

Word:
Definition:

My Glossary

Word:
Definition:

Word:
Definition:

Word:
Definition:

Word:
Definition:

Word:
Definition:

Word:
Definition:

My Glossary

Word:
Definition:

Word:
Definition:

Word:
Definition:

Word:
Definition:

Word:
Definition:

Word:
Definition:

My Glossary

Word:
Definition:

Word:
Definition:

Word:
Definition:

Word:
Definition:

Word:
Definition:

Word:
Definition:

STEAM Discussion Questions
Alexander Wetmore

Search for primary source documents here: http://collections.si.edu/search/index.htm

SCIENCE In Wetmore's journals, he listed the birds he saw each month. In a bird identification book, look up some of these birds he saw in February 1904: blue jay, Hairy Woodpecker, English sparrow, white-breasted nuthatch, chickadee, downy woodpecker, crow, great horned owl, junco, tree sparrow, ruffed grouse, barred owl, screech owl, bobwhite, American goldfinch, pine grosbeak, prairie horned lark, brown creeper, red poll, evening grosbeak, American goshawk, Acadian owl, Bohemian waxwing, American crossbill. Which of these birds, if any, live in your area?

Discussion: Keep track of the birds you see in your area for a week or a month. Discuss the common birds and their characteristics.

TECHNOLOGY Wetmore collected many specimens. This means he had to preserve the bird or animal bodies. He chose to dry many of the bird skins.

Discussion: In the mid 1900s, it was acceptable to dry bird skins in the sunlight. How would scientists today preserve birds and bird skins?

ENGINEERING As an ornithologist, or bird scientist, Wetmore needed good binoculars to see the birds he studied.

Discussion: How do binoculars solve a problem for the scientists? What other tools could solve the same problem?

Research a common bird in your area. What kind of bird house would that bird species need? Think about overall size, the diameter/shape of the entrance, nesting area, etc. If you have time, build a bird house, install it and observe birds.

ART Wetmore's first published paper, written at age 13, "My Experience with a Red-headed Woodpecker," appeared in Bird-Lore in 1900. The article is reproduced on pp. 148-149.

Discussion: Read and discuss Wetmore's narrative essay. What facts does he present? How does he use his own experiences in this essay? What observation surprised him in paragraph 3? How did the woodpecker hide his acorns? Wetmore includes information on the sounds of the woodpecker. How does that help make the narrative work? In the next-to-last paragraph, Wetmore mentions his notebook. How did it help him get the facts right in his story?

MATH When Wetmore listed birds sighted in a single month, he was using basic arithmetic. Sometimes just counting the number of species sighted in a month is enough. Since 1900, the Audubon Society has conducted a count of birds each December. Read more about it here: http://www.audubon.org/conservation/science/christmas-bird-count

Discussion: What can you learn about the health of birds just by counting the number of species each year? If possible, participate in the next Christmas bird count.

Martin H. Moynihan

Search for primary source documents here: http://collections.si.edu/search/index.htm

SCIENCE One of Moynihan's constant questions was about the difference in tropical and temperate animals.

Discussion: How does climate affect what animals survive in a certain area?

TECHNOLOGY To study squids, scientists use scuba gear. SCUBA stands for Self-Contained Underwater Breathing Apparatus.

Discussion: People need air to breathe. Besides underwater, what are conditions where people need technology to breathe? (Space, medical problems, caves with poisonous gas, etc.)

ENGINEERING The Panama Canal was a huge engineering feat. The goal was to create a system of waterways and locks and dams to connect the Atlantic and Pacific Oceans. Study the building of the Panama Canal.

Discussion: Barro Colorado Island was created when a river was dammed to make Gatun Lake. Discuss why it became an important island for tropical forest research.

Underwater, Moynihan used waterproof paper and ink to take notes.

Discussion: Design a way to waterproof paper and ink. Or, design some other way to take notes underwater.

ART Moynihan's journals originally used a chronological order. He dated his observations, and the notes are in order by date. Later, he reorganized the journals by topic. He grouped observations according to the animal he was observing.

Discussion: Which type of organization is better, chronological or topical? When would you use one over the other? Take some notes that are in chronological order and rearrange them into a topical order. Does this make information on the topics easier to understand?

Moynihan also used labels in his drawings. Make an outline drawing of something. Add labels to give the viewer more information.

MATH Humans and squid both have symmetrical bodies, but in different ways.

Discussion: Use geometry appropriate for your grade level to discuss symmetry of different animals. Use simple shapes to make a model of an animal. Divide an animal shape into equal parts. Draw lines of symmetry for different animals.

William Healey Dall

Search for primary source documents here: http://collections.si.edu/search/index.htm

SCIENCE The Western Union Telegraph Expedition attempted to install telegraph poles and cables in the Arctic area to connect North America with Asia and Moscow, Russia. Scientists accompanied the expedition to study flora and fauna (plants and animals). While the telegraph efforts failed, scientists learned much about the area.

Discussion: Discuss some things scientists might pay attention to as they travel.

Dall was a malacologist (mollusk scientist). Look at mollusks and discuss how a scientist would study them.

TECHNOLOGY In 1867, the telegraph was the cutting edge communication technology.

Discussion: Compare the telegraph with communication devices today.

ENGINEERING The plan to install telegraph poles and cables was a huge engineering problem. In 1858, the first trans-Atlantic underwater cables connected North America and England, and a permanent cable was completed in 1866.

Discussion: Compare the engineering problems of laying underwater cables with building a telegraph line in the Arctic. For example, to install one telegraph pole, workers had to build fires to thaw the ground and then dig the hole for the pole.

ART On Thursday, July 18, 1867, Dall wrote and drew about pottery made by natives. Here's what he wrote: "See for the first time pottery made by Indians and the only pottery made in the country. Very large pots of canteen holding from five gallons to half a gallon and a rude cup with two lumps of clay as handles; they are ornamented with dots, lines and crosses cut in the clay when soft and are baked in the sun and then in the fire. When in the fire are kept turning round and round like a joint of meat when roasting. They were of a dark brown, blackened by frost and use and pretty regular in shape." (Smithsonian Institution Archives. Diary

July 14-November 30, 1867 by William H. Dall. SIA2015-010202)

Discussion: What does the art add to your understanding of the pottery? What does the text add? Why is it important to both draw and write about something?

MATH Dall drew many maps on his travels.

Discussion: What math is needed to draw accurate maps. How would today's GPS (Global Positioning System) make map making easier?

Joseph Nelson Rose

Search for primary source documents here: http://collections.si.edu/search/index.htm

SCIENCE Rose and Britton's book about cacti was important because they changed the ideas about how to classify plants. Study the system of classifying plants: kingdom, phylum, class, order, family, genus, and species.

Discussion: Why do scientists use classification systems?

TECHNOLOGY Plant presses are made with simple technology. Research how a plant press works. An herbarium is any collection of dried plants that are arranged in some order. Make plant presses and create a classroom herbarium.

Discussion: For the classroom herbarium, how will you organize the specimens?

ENGINEERING Rose preserved plants by pressing or drying them. Also see how Wetmore preserved bird skins on page 15.

Discussion: Design a system for preserving and shipping specimens. Choose different types of specimens such as dinosaur bones, grasses, fossils, and birds. How would the preservation and shipping differ for each specimen?

ART Use dried flower petals to make a work of art.

Discussion: Discuss unusual materials to make a piece of art?

MATH Rose used geometric descriptions of plants: oblong and elliptical.

Discussion: What geometric shapes would help describe plants? Practice describing plants using geometric shapes.

William and Lucile Mann

Search for primary source documents here: http://collections.si.edu/search/index.htm

SCIENCE Some animals are almost extinct. They only live in captivity, especially in zoos. A zoo can be an important part of saving a species from extinction. However, animal rights activists think that zoos shouldn't cage animals.

Discussion: What are the pros and cons of zoos?

TECHNOLOGY Study the function of different body parts. Read about scientists who design 3-D printed body parts for injured animals.

Discussion: If you want to design a 3-D printed part, what do you need to know about how a body part works?

ENGINEERING Choose an animal and research its food and habitat needs. Design a zoo habitat for the animal that meets its needs.

Discussion: How do you find out an animal's food and habitat needs?

ART Lucile Mann liked to write and made time every day to write. Keep a diary for a week or a month, trying to write something every day.

Discussion: How do writers find time to write? What can you do to make it easier to write every day?

MATH Research how much food a certain animal eats in one day. Calculate how much food it would need for a week? A month? A year?

Discussion: How many days in a week? A month? A year? How do you calculate how much food is needed for a week? A month? A year?

Fred Soper

Search for primary source documents here: https://profiles.nlm.nih.gov/VV/

SCIENCE Some diseases are carried by mosquitoes. Study the life cycle of mosquitoes and how they carry diseases.

Discussion: Compare how diseases are transmitted among humans.

TECHNOLOGY Soper used a man-made insecticide to help kill mosquitoes. However, scientists later learned that DDT was dangerous for the environment. Research the history of DDT to understand why it was banned by the Environmental Protection Agency.

Discussion: Man-made chemicals hold great promise for many uses. However, the chemicals are new to the environment. It might take years to understand how a chemical would affect the environment. What safeguards would help protect the environment, and yet, let people use new chemicals for a certain purpose.

ENGINEERING Soper's work in public health focused on sanitation. Study or visit the sanitation, water or other departments of your city.

Discussion: How does a city sanitation department help to keep people healthy?

ART Soper was amused that a city would celebrate the opening of a latrine. But the party helped people to understand the importance of latrines. Plan a public health announcement. Make a poster, videotape a message, or plan a party to help people understand a public health issue.

Discussion: Television stations are required to offer free time for "public service announcements." Talk about why this is important.

MATH Decisions about public health issues often require statistics. For example, the Rockefeller Institution decided to work on the hookworm problem because of statistics. In 1910, 40% of Americans in the South had hookworms. Statistics is the science of collecting and analyzing large sets of numbers.

Discussion: Create a simple line graph that shows the birth month of students in your class. How did you collect the information? How did you show the information? Gather and discuss other simple statistics for your class. Here are some ideas: number of siblings, number of pets, or shoe sizes.

Mary Agnes Chase

Search for primary source documents here: http://collections.si.edu/search/index.htm

SCIENCE Chase used a microscope to help her draw more exact images of plants. Use a magnifying glass to study plant anatomy.

Discussion: Compare what you can see with a microscope and a telescope. When would you use each in scientific observation?

TECHNOLOGY Chase typed all of her journals.

Discussion: How has technology changed the way we store field notes? Discuss the pros and cons of using computers for taking field notes.

ENGINEERING In her 1922 book, *The First Book on Grasses, the Structure of Grasses Explained for Beginners*, Chase wrote, "Of all plants, grasses are by far the most important to man. The grains of wheat, barley, rye, oats, rice, corn sorghum, and millet form the staple food of the greater part of mankind. . .The grains are also the sources of starch, alcohol, and glucose." Discuss the importance of grasses to human civilizations. Discuss the engineering solutions that allow the grains to be used for food. In other words, how are seeds of grasses, such as wheat, turned into flour?

Discussion: A mill is a place with machinery for grinding grain into flour and other cereal products. Design or make a small mill. It could grind coffee, oats, corn or other plants.

ART Chase began her career as a botanical illustrations. Study other botanical artists. This website lists famous botanical artists worldwide: http://www.botanicalartandartists.com/artists.html

Botanical illustrations often show an idealized plant, and includes the whole plant. They may also choose to show the plant in various stages, such as a flower bud and a fully open flower. As you look at the illustrations, pay attention to how much of the plant is shown, and in what context. Do the illustrations show contrasts and comparisons?

Discussion: What is the advantage of having a botanical illustrator as well as a photographer? What can an illustrator do that a photographer can't? What can a photographer do that an illustrator can't?

Write a narrative of a trip.

Discussion: Discuss the importance of using time words as you write a narrative. Talk about the difference in revising and editing. Revising is when you rethink the story and make big changes. Editing is correcting spelling and grammar.

MATH The Food and Agriculture Organization (http://www.fao.org/docrep/008/y8344e/y8344e05.htm) estimates that about 40% of the world (excluding Antarctica and Greenland) is grassland. How much of the United States is grassland? How would you figure out the percentage of grassland in your country?

Discussion: How do you figure percentages? How could you chart, diagram or otherwise display the information in an interesting and informative way?

Donald S. Erdman

Search for primary source documents here: http://collections.si.edu/search/index.htm

SCIENCE In August 1948, Erdman recorded an open water temperature of $39C°$ (approx. $104F°$). Locate Tarut Bay or Ras Tanura in the Persian Gulf on a map or on an app such as Google Earth.

Discussion: How does nearby land affect the temperature of a body of water?

TECHNOLOGY Fishing is a popular sport and hobby. Erdman studied the blue marlin partly to make sure the sport of fishing would remain strong.

Discussion: Talk about the technology used to do sport fishing. What kind of equipment is used and how has that changed?

ENGINEERING The Arab fisherman designed an unusual way to catch fish. They used the natural process of tides to trap fish in a V-shape corral.

Discussion: Design a way to capture fish by taking advantage of the tides or other natural events.

ART See Erdman's fish in full color here: https://www.flickr.com/photos/smithsonian/albums/72157651391850445

Use color to make a piece of art. But don't use color just to use color. Erdman used color because the fish turned brown or black when they were preserved. Use color to give more information about something.

Discussion: Study the color wheel and plan the colors you will use.

MATH Discuss thermometers and the difference in Celsius and Fahrenheit scales. Practice converting one scale to the other.

Discussion: Scientists often record weather temperatures in their notebooks. Talk about times when the temperature could affect the results of a science experiment.

Robert E. "Bob" Silberglied

Search for primary source documents here: http://collections.si.edu/search/index.htm

SCIENCE Silberglied studied how butterflies communicate using ultraviolet patterns.

Discussion: How do butterfly eyes see ultraviolet? How could you determine if they could see ultraviolet patterns?

TECHNOLOGY Silberglied's work depended heavily on black lights and UV-lights.

Discussion: Discuss the technology of making ultraviolet lights. How are black lights and ultraviolet lights used today? For example, the security thread of a US$20 bill glows green under black light as a safeguard against counterfeiting.

ENGINEERING Photography is the capture of light to create an image. From his early days on the Ant Safari, Silberglied was interested in ultraviolet. How would you take pictures using ultraviolet light? Photography can also be used to take pictures using a microscope to magnify something. The electron microscope, invented in the late 1920s and 1930s, meant scientists could now take pictures of even smaller things. On the other hand, aerial (think drones!) or satellite images pull back to make things smaller, but give a wider perspective. Discuss how photography can be used to help scientists study something.

ART Silberglied drew many illustrations, but usually drew an outline and labeled things. Discuss the difference in detailed photos and outline drawings with labels. Study his map and the key that he included.

Discussion: How does the key help explain the map?

MATH In Silberglied's notebooks, he often counted the daily catch of insects and bugs, kept track of the vehicle's mileage and other simple math tasks.

Discussion: Talk about the importance of recording exact figures in a notebook.

Watson M. Perrygo

Search for primary source documents here: http://collections.si.edu/search/index.htm

SCIENCE Taxidermy is the process of preserving an animal skin. Perrygo talked about the difficulty of preserving the hippopotamus's skin because it was so large and because it was a water animal. A 2010 report from taxidermists said it was one of the best preserved specimens. Even after 80 years, it was in good shape.

Discussion: Preservation is the opposite of the process of decay. Study the process of decay. Why is decay a helpful process? For example, leaves in a compost heap decay and become new soil.

TECHNOLOGY Visit a museum or taxidermist. Notice the animals that have been preserved with taxidermy.

Discussion: What technology is needed for taxidermy?

ENGINEERING Design a museum display.

Discussion: What are some ways to display information?

ART Taxidermy tries to preserve animal skins. However, fossils don't have skin. For dinosaurs and other fossilized animals, no one really knows what they looked like. Their skin might have been any color. Draw dinosaurs and make their skin unusual colors.

Discussion: Is it possible that the Tyrannosaurus Rex had pink skin?

MATH One of the most important exhibits at the Smithsonian Museum is the Fénykövi Elephant, which is displayed in the rotunda. The elephant was donated to the museum by Josef J. Fénykövi. The hide or skin of the elephant weighed two tons. Read about the elephant here:

http://naturalhistory.si.edu/onehundredyears/featured_objects/Fenykovi_elephant.html

Notice in the drawing on the next page, the elephant's measurements are given in many different directions. They used both lengths and circumferences. Review the procedures for finding length and circumference.

Measure a pet (or stuffed animal). Using the elephant drawing as an example, decide what measurements to take of a pet. For example, you may want to measure the length of a tail, leg, ear or nose. You could measure the circumference of the pet's chest, arm, leg, or head. Use a tape measure to measure. Draw an outline of the pet and label it with measurements taken.

Discussion: Measurements add information to an outline drawing. How do the measurements help you understand the size of the animal? Would you understand the animal as well without the measurements?

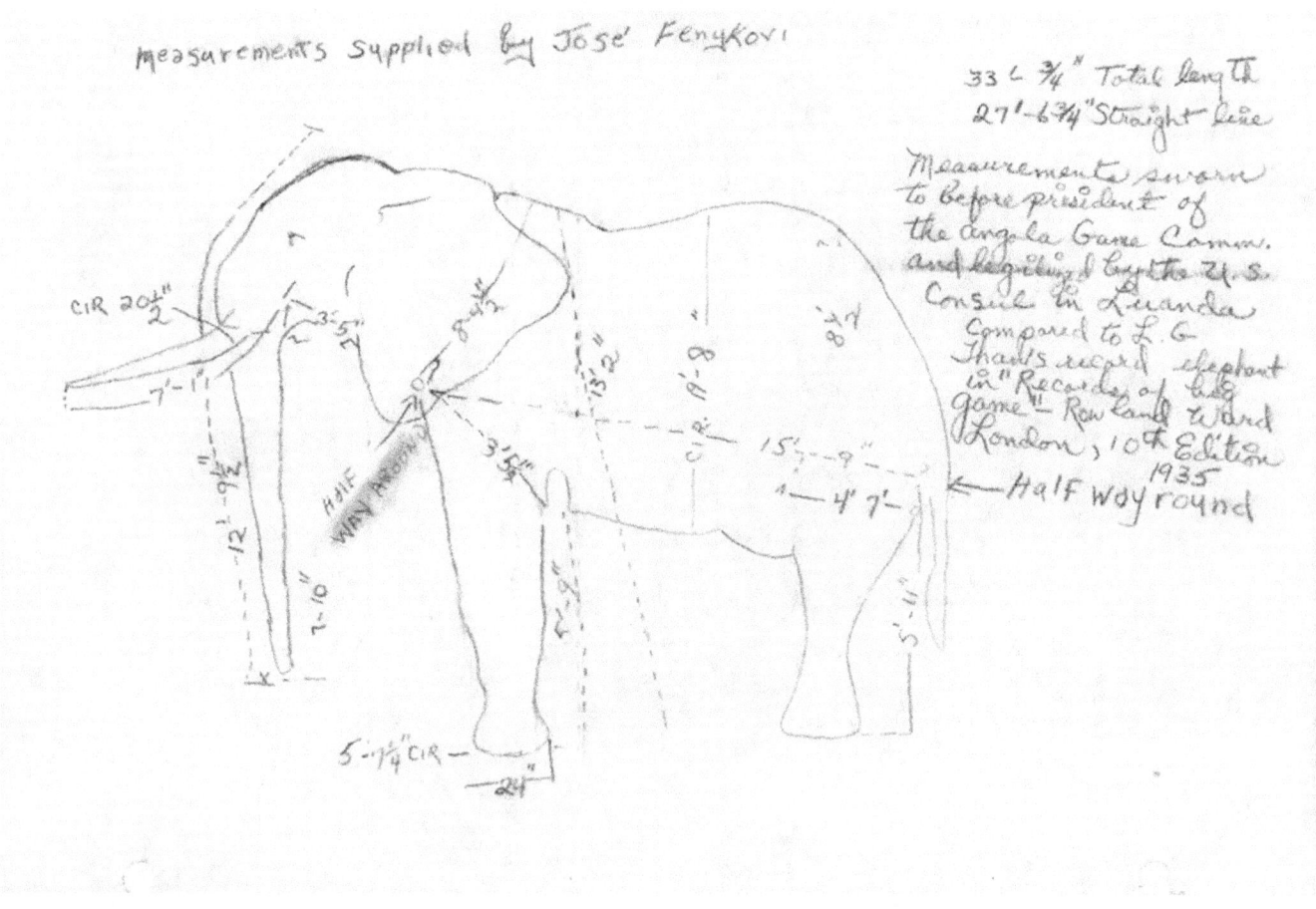

Reproduction of p. 155-6 *Bird Lore* magazine, 1900. This was Alick Wetmore's first published article at age 15. Public domain.

For Young Observers

My Experience with a Red-headed Woodpecker

BY ALICK WETMORE (age, 13 years), North Freedom, Wis.

THE first time that I saw the subject of this sketch was on Sunday, October 8, 1899. As I was going along a ravine on that day, I heard a loud, tree-toad-like *ker-r-r-ruck* coming from the top of a tall dead stub. I looked up and soon saw that the owner of the voice was a young Red-headed Woodpecker. His (?) head was a dusky color. He would stick his head around the tree and, after giving the note, dodge back. I thought I would keep a sharp eye on him, and a little while afterward I was rewarded by seeing him get an acorn from a small oak. He seemed to be storing acorns up for winter in holes and crannies.

Once he lit on an oak limb that would not bear him, and it swung until he hung back down, but he got his acorn. While he was flying off, a little Junco seemed to think that he was trespassing and flew at him in a rage and made him get out of the way. I went to a stump nearby and got an acorn and found that it was whole. A few marks on the shell showed where he had hammered it into the crevice. He always seemed to go to the same tree for his acorns.

I laid down on the bank of the ravine close to the tree in the sun to watch him, but he was suspicious and would not come near at first. I was rather surprised to see that he could easily go down a tree backwards, lifting his tail and, after hopping down, falling back onto it. Everywhere he went, he expressed, in vigorous notes, his disgust at having me around.

The stub he liked best was very tall and had a crack in it near the top, and into this crack he hammered, with his shiny white bill, all the acorns that he possibly could. Some of them he cracked in two and then put them in the crack. One fragment he dropped as he lighted. He was after it quick as a flash, and chased it so near the ground that I thought he would dash himself onto it and be killed, but he turned up just before he reached it and flew off without the acorn.

In a cornfield a short distance away I found some nubbins for him. While I was looking for a place to put them up, I found a hole with sixteen acorns in it. He had put them there, for I could see the marks of his bill on them and around the edges of the hole were a few small dark gray feathers. He had hidden the acorns by putting pieces of bark over them. I then went back to where he was and saw him drinking water, like a chicken, out of the brooklet. After returning from a short walk, I saw him carrying a large piece of bark to put over the acorns that I had uncovered. He started from the base of his stub, but as the bark was nearly as large as he was he could not carry it and was forced to drop it. As it was then nearly dark, I had to go home without learning where he stayed nights, and which, indeed, I never found out.

The next Sunday, the 16th of October, I did not have much time. When I reached the ravine he was catching insects. He was in the top of a tree and would fly out after the insects at they flew by but, growing tried of this, he went to the ground after an acorn. When I went to the hole in which I had found the sixteen acorns before, I now took out forty-five.

Sunday, November 19, I thought I would pay my Red-head a visit. As I did not see him for about fifteen minutes, I thought that some wandering hunter had killed him; but while looking around I heard a welcome *ker-r-r-ruck*, and there he was on his favorite stub. After taking a look at me, he flew down for a drink, with a loud note before he left the stub and shorter ones in between drinks to call attention, and well he might! His somber head had turned red since I had seen him last. The color was a little dark in places, but was fine all the same.

I next saw him on Sunday, November 26. I had gone to my usual place of study and was watching some Pine Siskins when he appeared. He was rather cross, for he chased a Tree Sparrow until it took refuge in a thick, bushy thorn-apple tree. Then he watched until it came out and took after it agan. I watched him sunning himself—for it was quite warm—and then went over to the hole in which I had found so many acorns. It was empty, and a number of shells were scattered around the foot of the tree.

From my note-book I see that the date of my next visit was Sunday, December 3. It was cold and snowing quite hard. I put on my overcoat and went down to see him. I may have wanted to see him, but he was evidently afraid of that big black thing in the fence-corner. He scolded and bobbed as though crazy till a pair of Blue Jays lighted in the tree. He was afraid of them and went around to the other side of the trunk and kept still until they left.

On Monday February 12, I saw him last. He was across the river from the ravine in a tree after acorns.

I know that he is still here and alive, and I intend to watch him in the spring when he sets up housekeeping.

Photo Permissions & References

Why a Science Notebook

1. Parilla, Lesley. In the Field for the Holiday. Field Book Project Website. http://nmnh.typepad.com/fieldbooks/2011/12/in-the-field-for-the-holidays.html. Accessed August 30, 2016
2. http://naturalhistory.si.edu/rc/fieldbooks/
3. https://www.nlm.nih.gov/
4. http://www.computerhistory.org/
5. Wetmore, Alexander. 1894 notebook. Smithsonian Institution: SIA RU007006.
6. Siberglied, Robert. Field notes, Mexico, July-August, 1965. Smithsonian Institution: SIA2015-000087.
7. Mann, Lucille. Diary of the National Geographic Society-Smithsonian Institution Expedition to the Dutch East Indies, 1937. Smithsonian Institution: SIA2014-07220.
8. Silberglied, Robert. op.cit

How to Use This Notebook

9. Fulton, Lori and Emily Poeltler. Developing A Scientific Argument. Science and Children. Summer, 2013, pp. 30-35

Alexander Wetmore

Quote from 1895 journal. Field notes, Florida, 1894-1895 : his first recorded natural history observations taken at the age of eight. Smithsonian Institution. Archives. Accession #: SIA RU007006 Box 131 Folder 1

Patuxent Wildlife Research Center website. Alexander Wetmore. http://www.pwrc.usgs.gov/resshow/perry/bios/wetmorealexander.htm. Accessed on August 18, 2016.

Smithsonian National Museum of Natural History website. Celebrating 100 Years. Alexander Wetmore: Ornithologist and Sixth Secretary of the Smithsonian. http://naturalhistory.si.edu/onehundredyears/profiles/Alexander_Wetmore.html. Accessed on August 18, 2016.

Smithsonian Olmec Legacy website. Archaeologists & Scholars. Alexander Wetmore 1886-1978. http://anthropology.si.edu/olmec/english/archaeologists/wetmore.htm. Accessed on August 18, 2016.

Photo Permissons:
Wetmore Riding Horse. Smithsonian Institution Archive. Image # SIA2016-000427.
Wetmore drying bird skins. Smithsonian Institution Archive. Image # SIA2008-2949.
Wetmore, age 15, in his room. Smithsonian Institution Archive. Image # MNH-17021.
Wetmore's February, 1904 Bird List. Smithsonian Institution Archive. Image # SIA2015-010126.

Martin Moynihan

Lawton, Rebecca. Consistently Lovely: The Exceptional Field Notes of Martin M. Moynihan. Hakai Magazine website, October 23, 2015. https://www.hakaimagazine.com/article-short/consistently-lovely-exceptional-field-notes-martin-h-moynihan Accessed on August 24, 2016.

Parilla, Lesley. A Behaviorist in Panama. Smithsonian Field Book Project website. January 4, 2012. http://nmnh.typepad.com/fieldbooks/2012/01/a-behaviorist-in-panama-.html. Accessed August 24, 2016.

Saxon, Wolfgang. (December 15, 1996). "Martin H. Moynihan, 68, an Authority on Animal Behavior." New York Times access on December 1, 2011 at http://www.nytimes.com/1996/12/15/world/martin-h-moynihan-68-an-authority-on-animal-behavior.html

Smith, Neal Griffith. (July 1998). "In Memoriam: Martin Humphrey Moynihan, 1928-1996." American Ornithologist's Union. Published by University of California Press. Accessed December 1, 2011 at http://www.jstor.org/stable/4089423

Smithsonian Tropical Research Institute. "Highlights: Important events in the history of Barro Colorado." Accessed December 1, 2011 at http://www.stri.si.edu/sites/forest_speaks/english/about_forest_speaks/history/index.html

Photo Permissions:
Moynihan in window. Smithsonian Institution Archive. Image # SIA2008-2981.
Squids notebook. Smithsonian Institution Archive. Image # SIA2008-2981.

Notes with Primate. Smithsonian Institution Archive. Image # SIA2014-02581.
Bird with notes. Smithsonian Institution Archive. Image # SIA2014-03311.

William Healey Dall

Hunter, Cathy. William Dall: National Geographic founder and Pioneer of Alaskan Exploration. June 15, 2012. National Geographic Website. http://voices.nationalgeographic.com/2012/06/15/william-dall-national-geographic-founder-and-pioneer-of-alaskan-exploration/ Accessed on August 24, 2016.

PBS Website. Harriman Expedition Retraced. http://www.pbs.org/harriman/1899/1899_part/participantdall.html Accessed on August 24, 2016

Smithsonian Institution Archives. William Healey Dall: Alaskan Explorer. http://siarchives.si.edu/history/exhibits/stories/william-healey-dall-alaskan-explorer Accessed on August 24, 2016.

Wording, W.P. William Healey Dall 1845-1927: A Biographical Memoir. Washington, D.C." National Academy of Sciences 1958.

Photo Permissions:
Dall in Western Union Telegraph Expedition uniform. Smithsonian Institution Archive. Image # 2006-18833.
Drawing of Native Woman. Diary July 14 - November 30, 1867- William Healy Dall. Smithsonian Institution Archive. Image # SIA2015-010193.
Native Pottery; Diary July 14 - November 30, 1867- William Healy Dall. Smithsonian Institution Archive. Image # SIA2015-010202.
Map of Alaskan rivers; Diary July 14 - November 30, 1867- William Healy Dall. Smithsonian Institution Archive. Image # SIA2015-010188.

Joseph Nelson Rose

Memorial: Joseph Nelson Rose, Science. 15 Jun 1928: Vol. 67, Issue 1746, pp. 598-599. DOI: 10.1126/sciencc.67.1746.598.

Padilla, Lesley. Get to Know Joseph Nelson Rose. April 22, 2014. Field Book Project website. http://nmnh.typepad.com/fieldbooks/2014/04/joseph-nelson-rose.html Accessed on August 24, 2016.

Socha, Aaron, M. Joseph Nelson Rose and "The Cactaceae." From Areoles to Zygocactus: An Evolutionary Masterpiece. New York Botanical Garden website: http://www.nybg.org/bsci/herb/cactaceae1.html#Rose Accessed on August 24, 2016.

Photo Permissions:
Headshot at age 23. Rose JN PD_055_SS09 Image courtesy of the Robert T. Ramsay, Jr. Archival Center at Wabash College.
Cactus greenhouse record- Jacob Nelson Rose. Smithsonian Institution Archive. Image # SIA2013-09492.
Bartschella schumannii Britton & Rose. Collected by J. N. Nelson in Northwest Mexico, on March 23, 1911. US-638432. Courtesy of U.S. National Herbarium, Smithsonian Institution

Lucile and William Mann

About Zoos website. VIPsZoos: William M. Mann. http://aboutzoos.info/zoos/vips-zoos/75-william-m-mann#biography Accessed August 24, 2016.

Henson, Pamela M. A Life on the Wild Side: Lucile Quarry Mann. The Bigger Picture Blog. Smithsonian Institution Archives website. March 15, 2011. http://siarchives.si.edu/blog/life-wild-side-lucile-quarry-mann Accessed August 24, 2016.

Smithsonian Institution Archives, SIA RU009513, Oral History Interviews with Lucile Quarry Mann 1977.
Smithsonian Institution Archives, Record Unit 7293, , William M. Mann and Lucile Quarry Mann Papers http://siarchives.si.edu/collections/siris_arc_217450 Accessed August 24, 2016.

Photo Permissions:
Lucile Mann Feeding cub, Babette. Smithsonian Institution Archive. Image # 79-11710.
Manns in British Guiana. Smithsonian Institution Archive. Image # 79-11666.
L.Mann diary in Singapore. Smithsonian Institution Archive. Image # SIA2014-07220.

Fred Soper

The Fred L. Soper Papers. Profiles in Science. National Library of Medicine. https://profiles.nlm.nih.gov/VV/ Accessed on August 24, 2016.

Gladwell, Malcom. The Mosquito Killer, Originally in the New Yorker Magazine. July 2, 2001. Archived at http://gladwell.com/the-mosquito-killer/ Accessed on August 24, 2016.

The Rockefeller Archive Center website. Eradicating Hookworm: 100 Years Rockefeller Foundation. http://rockefeller100.org/exhibits/show/health/eradicating-hookworm Accessed on August 24, 2016.

Photo Permissions:
Fred Soper. Courtesy of National Library of Medicine. ID# VVBBCB.
Boy with hook and tape worms. Courtesy of the National Library of Medicine. ID# VVBBJC. Rockefeller Archives.
Excerpt from Fred Soper's diary. Courtesy of National Library of Medicine, ID# VVBBGL.

Mary Agnes Chase

Boyd, Teresa. The Other Side of Mary Agnes Chase. Smithsonian Institution Archives website. March 19, 2015. http://siarchives.si.edu/blog/other-side-mary-agnes-chase Accessed on August 24, 2016.

Chase, Mary Agnes, The First Book of Grasses. New York: Macmillan. http://biodiversitylibrary.org/item/61054#page/7/mode/1up. Accessed August 24, 2016.

Chase, Agnes. Database of Scientific Illustrators, 1450-1950 . http://www.uni-stuttgart.de/hi/gnt/dsi2/index.php?table_name=dsi&function=details&where_field=id&where_value=6416. Accessed August 24, 2016.

Mary Agnes Chase (1869-1963). United State National Arboretum website. http://www.usna.usda.gov/Education/Chase.pdf Accessed August 24, 2016.

Persons, Collections and Topics: Hitchcock-Chase. Hunt Institute for Botanical Documentation. http://www.huntbotanical.org/art/show.php?5 Accessed on August 24, 2016.

Photo Permissions:
Chase on Mt.Top with Donna Maria. Smithsonian Institution Archive. Image # SIA2009-4227.
1929 Chase Journel of Mt. Climb. Smithsonian Institution Archive. Image # SIA2015-003221.
First Book of Grasses, c.1922 Macmillian. p. 28 Public Domain.

Donald S. Erdman

Civilian Public Service Website. CPS Worker 002669 - Erdman, Donald. http://civilianpublicservice.org/workers/2669 Accessed on August 24, 2016.

Donald S. Erdman Papers, 1948. Smithsonian Institution Archives. http://siarchives.si.edu/collections/siris_arc_217583 Accessed on August 14, 2016.

Edman, Donald S. Inland Game Fishes of Puerto Rico. Commonwealth of Puerto Rico, Department of Agriculture, 1972.

Photo Permissions:
Portrait 1949 in Saudi. Assumed to be Erdman. Smithsonian Institution Archive. Image # SIA2014-06857.
Donald S. Erdman Papers, 1948. Spangled Emperor. Smithsonian Institution Archive Image # SIA2014-06697.
Palm State Trap. Smithsonian Institution Archive. Image # SIA2014-06867.

Robert Silberglied

Carpenter, Frank M. Psyche: A Journal of Entymology. Dedication: Robert E. Silberglied. Psyche 88:197-198, 1981. http://groups.csail.mit.edu/mac/projects/psyche/88/88-197.html. Accessed on August 24, 2016.

Hunter, Emily. Butterfly Vision: Robert E. Silberglied's Photographic Explorations. Field Book Project website. September 21, 2012. http://nmnh.typepad.com/fieldbooks/2012/09/butterfly-vision-robert-e-silberglieds-photographic-explorations.html Accessed on August 24, 2016.

Smithsonian Tropical Research Institute: Panama. New BCI species named after STRI's Robert "Bob" Silberglied (1946-1982). May 19, 2008. http://stri.si.edu/english/about_stri/headline_news/news/article.php?id=800. Accessed on August 24, 2016.

Smithsonian Institution Archives, Record Unit 7316, Silberglied, Robert Elliot, Robert Elliot Silberglied Papers http://sova.si.edu/record/Record%20Unit%207316 Accessed on August 24, 2016.

Yoffe, Emily. Bridge of Sighs. The Guardian, January 11, 2003. https://www.theguardian.com/theobserver/2003/jan/12/features.magazine27 Accessed on August 24, 2016.

Photo Permissions:
Silberglied sitting on Galapagos Tortoise. Photo courtesy of Frank J. Sulloway.
Silberglied journal, 1965 Mexico Trip and close-up of map key. Smithsonian Institution Archive. Image # SIA2015-000103.
BW Butterfly. Smithsonian Institution Archive. Image # SIA2012-7947.
Butterfly Wing Close-up. Smithsonian Institution Archive. Image # SIA2012-7946.

Watson M. Perrygo

Charles County Historical Society. Watson M. Perrygo - Historical Society of Charles County. Charles County Historical Society Website. charlescountyhistorical.org/Watson_Perrygo.doc Accessed on August 24, 2016.
Smithsonian Institution Archives, Record Unit 9516, Perrygo, Watson M. interviewee, Watson M. Perrygo Interviews. http://siarchives.si.edu/collections/siris_arc_217684 Accessed on August 24, 2016.
Starts, Siobhan. A Scientist and a Tinkerer - A Story in a Frame. National Museum of Natural history website. http://nmnh.typepad.com/100years/2010/08/a-scientist-and-a-tinkerer-a-story-in-a-frame.html August 10, 2010. Accessed August 24, 2016.
Perrygo, Bruce A. "The Colorful Life of a Smithsonian taxidermist." Washington Post, August 17, 2015. https://www.washingtonpost.com/opinions/the-colorful-life-of-a-smithsonian-taxidermist/015/08/17/8c61e916-4298-11e5-9f53-d1e3ddfd0cda_story.html?utm_term=.2a2daa451281.

Photo Permissions:
Perrygo with snake. Smithsonian Institution Archive. Image # 84-8050.
Watson M. Perrygo preparing specimens in Camp at Don Don, Haiti. Smithsonian Institution Archive. Image # SIA2008-2555.
Measurements with diagram of the Fenykovi Elephant. Smithsonian Institution Archive. Image # SIA2010-0605.

Read More About Using Science Notebooks

Ansberry, Karen and Emily Morgan. Words to the Wild, Teaching through Trade Books. Science and Children, November 2007
  Article: https://learningcenter.nsta.org/resource/?id=10.2505/4/sc07_045_03_14
  To accompany the article, the NSTA provides a sample rubric to assess science notebooks.
  Rubric: http://www.nsta.org/elementaryschool/connections/200711Rubric.pdf

Fulton, Lori and Brian Campbell. Science Notebooks: Writing About Inquiry. Portsmouth, NH: Heinemann 2014.
  Excellent book on the education theories and practices of using a science notebook in classrooms.

For more primary source documents, refer to The Field Book Project which began in 2010 as a joint initiative between the Smithsonian Institution National Museum of Natural History (NMNH), Smithsonian Libraries, and the Smithsonian Institution Archives. The project has cataloged more than 7,500 field books across eight departments within the Smithsonian and plans to add another 2,600 by 2018.